EARTH

MOB

REDUCE WASTE. SPEND LESS. BE SUSTAINABLE

PAVILION

ACKNOWLEDGMENTS

I would firstly love to thank Sophie Godwin for writing this book. You are
incredible, and I am grateful every day that you are part of the MOB team.
Thanks also to my wonderful publisher Pavilion for supporting this important idea.
Finally, thanks to OMSE for their work on the design, the brilliant Bill Rebholz on
illustration and to my amazing MOB team, for all your work behind the scenes.
Ben x

First published in the United Kingdom in 2020 by
Pavilion
43 Great Ormond Street
London
WC1N 3HZ

Copyright © Pavilion Books Company Ltd 2020
Text copyright © Mob Kitchen 2020

ISBN 978-1-91166-327-0

A CIP catalogue record for this book is available from the British Library.

10 9 8 7 6 5 4 3 2

Printed and bound by 1010 Printing International Ltd, China

www.pavilionbooks.com

Publisher: Helen Lewis
Editor: Cara Armstrong
Designers: Laura Russell and Hannah Naughton
Design Consultants: OMSE
Illustrations: Bill Rebholz
Production Controller: Phil Brown

CONTENTS

WELCOME ON BOARD MOB

We're over the moon that you're here to join us on this shared mission to becoming more sustainable. This book is about an issue very close to all of our hearts. It's about our future and our planet.

At MOB Kitchen we are striving to be better; and this means being kinder to our planet by using up everything in our fridge, cutting down on all waste and making more eco-conscious shopping decisions.

We're not going to lie; it's been a challenge. When we started our journey to becoming more sustainable, it was incredibly daunting. There was so much information to take on board.

The issue was that we wanted to be better, but we didn't really know where to start. There is so much noise about all the things we *should* be doing. But what should we focus on?

We had simple questions like which plastics are really recyclable? What do 'free range' and 'organic' actually mean? Can you really use a Parmesan rind in your cooking?

What we needed was a comprehensive, no-frills guide about how to be more sustainable in the kitchen – but it didn't exist. So, we've created it.

We know how much you care about our planet, and this is our attempt to help. *EARTH MOB* contains answers to all your food-related queries, and in this handbook we want to give you all the tools that you need to be the best, most ecologically-minded version of yourselves, without getting overwhelmed by the task.

Our future starts at home with all of the small choices we make every day around food. Let's do this.

TIME TO ACT

First things first, let's talk about what's happening to our planet. Even if across the world certain individuals and organizations are playing ignorant, the effects of climate change are already being felt.

To put it into perspective, each year over 20 million people are forced out of their homes because of climate change.[1] Be it devastating cyclones in Africa, deadly floods and landslides in South Asia, crop-failing droughts in Central America, the hottest summers on record in Europe or the heart-wrenching Australian bushfires, we can no longer ignore what is going on.

SO, HOW IS THIS HAPPENING?

TEMPERATURES RISE

↓ ↓ ↓

| Ice melts | Warmer atmosphere holds more water | Intense periods of drought |

↓ ↓ ↓ ↓

| Sea levels rise | Severe tropical storms and hurricanes | Flash flooding and landslides | Fires catch and spread more easily |

↓

| Land erosion |

Our ecosystems are being put to the test by increasingly extreme weather conditions. Nature is resilient, but with devastation happening on an increasing scale it needs a helping hand.

Anything we can do now to reduce our individual climate impact is crucially important and it DOES make a difference.

WE CAN ALL MAKE A DIFFERENCE

To be clear MOB – taking the full responsibility of climate change on yourself is simply not fair. In order for large-scale change to happen, governments around the world need to step up to the plate and change their policies. In the meantime, we all have a voice and we can use it to come together and make some noise.

Take Greta Thunberg. If you haven't already heard of her we are not sure what rock you have been living under for the past two years, but, to fill you in, this teenage Swedish climate change activist has been making waves around the world, spear-heading a global call for action.

One Friday in August 2018, Greta decided that instead of going to school she would start an environmental strike outside the Swedish parliament to protest against the lack of action on the climate crisis. Her three-week long initial protest gained publicity, and now every Friday students around the world follow in her footsteps striking for #fridaysforfuture.

Greta has Asperger's syndrome, which she considers a gift because it allows her to see the world and its climate crisis in black and white. She gives refreshingly honest speeches, challenges global leaders on their environmental policies and has sailed across the Atlantic on a zero-carbon yacht twice to attend a UN action summit. Greta is a leader and we should all be doing what we can to follow her example.

"NO ONE IS TOO
SMALL TO MAKE A DIFFERENCE"

ENVIRONMENTAL

THE BAD NEWS

→ The Earth is currently on track for a **2.9–3.4°C** (37.2–38°F) increase in temperature by 2100. Experts warn that this increase must be kept below **1.5°C** (34.7°F) to avoid devastating impacts.[2]

→ The United Nations reports that **1 million** species of plants and animals are at risk from **extinction**.[3]

→ The world's food system is responsible for at least ⅓ of all greenhouse gas emissions.[4]

→ **One football pitch** of forest is destroyed every **second**.[5]

→ The rate of Antarctic ice loss has **tripled** in the last decade.[6]

→ Climate change is already affecting food security.

→ The World Health Organization (WHO) estimates that **4.2 million** people die globally every year as a result of air pollution.[7]

→ In 2019 the Earth Overshoot Day was **29 July**. This means that by 29 July we had used up all of that year's regenerative resources.[8]

HEADLINES

THE GOOD NEWS

→ An estimated **6.6 million** people across **185 countries** took part in the global climate strikes in September 2019.[9]

→ **1 in 3** people are now shopping more consciously.[10]

→ Costa Rica has one of the most ambitious climate change plans to completely decarbonize its economy by 2050. **98%** of its electricity already comes from **renewable resources.**[11]

→ **Generation Z** (those born between 1995 and 2012) have the highest levels of social responsibility and are the most globally mobilized around causes they believe in.[12]

→ There are **170+** global organizations addressing climate change.[13]

→ Stricter restrictions have been placed on **single-use plastic** around the world.

→ The European Green Deal aims to make Europe the first **climate-neutral** continent by 2050.[14]

→ Everyone is **talking** about climate change – **even the Pope.**[15]

→ More and more trees are being planted. Ethiopia planted more than **350 million trees in just one day!**[16]

MOB, IT'S TIME FOR CHANGE.

First things first, and that's getting the kit for sustainability. These are our reusable staples →

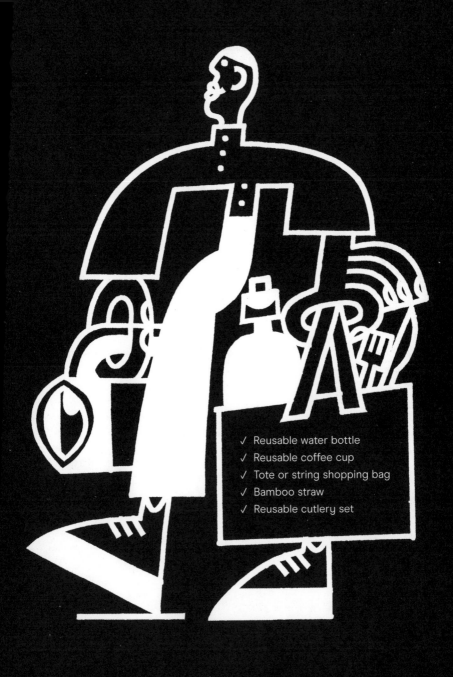

✓ Reusable water bottle
✓ Reusable coffee cup
✓ Tote or string shopping bag
✓ Bamboo straw
✓ Reusable cutlery set

SEASONALITY

Part of becoming more sustainable in the kitchen is understanding where our food comes from. When nearly everything we buy is pre-washed, pre-packaged and sometimes even pre-cut, it is easy to forget that our fruit and veg have been grown, loved and tended to by another human being.

To put it differently — if you had the time and the patience of a saint, you could even grow a lemon tree from a pip —

that's if you didn't mind waiting up to 15 years for it to produce any fruit.

So MOB, to help make sure that we don't absent-mindedly throw out half a courgette that has been lurking at the back of our fridge, we've got to appreciate the effort that goes into growing it in the first place.

That starts with a shout-out to the farmers who dedicate their lives to producing food for our plates.

FARMING

An estimated 95% of our food is produced using soil.[17] It is the foundation of our food system and maintaining its health is vital.

Soil is its own living ecosystem, teeming with bugs and all sorts of life. It converts dead matter into nutrients, acts as a water filter, controls diseases and even mitigates against climate change by storing carbon – we NEED to care for it.

Healthy soil = lots of good-quality fruit and veg = happy humans.

Unfortunately, over the past 50 years the soil has taken a battering. In order to feed an ever-increasing population, we have had to find ways of producing food faster. This means spraying our crops with harsh chemicals and repeatedly using the same piece of land to grow crops year in, year out.

These industrialized farming processes harm the soil, repeatedly stripping it of its natural nutrients, without allowing time for recovery. Luckily, there are some fantastic farmers out there who are trying to save our soil.

These farmers are using more natural methods that promote and maintain our plants and wildlife. Growing different crops alongside one another is one of the best ways to keep our soil happy, which is good news because we love diversity and so do plants.

So, what can the MOB do to help?

→ Try different types of fruit and veg. There are over 30,000 edible plant species in the world, and we currently only eat 150–200 of them![18]
→ Support your local farmers. If you hear of a local market happening then go along – it's a great way to get quality produce at a good price.
→ Eat more seasonally. This means trying not to buy asparagus in winter or parsnips in the summertime.

That last one is an absolute game changer. Eating seasonally means our food is fresher, more nutritious and above all **MORE DELICIOUS**. And don't worry if you don't know what's in season when, the following pages contain a guide to eating for the seasons.

SPRING

Veg:
→ Asparagus
→ Cauliflower
→ Jersey Royal new potatoes
→ Peas
→ Purple sprouting broccoli
→ Spinach
→ Spring greens
→ Spring onions
→ Radishes
→ Watercress

Fruit:
→ Elderflower
→ Rhubarb*

Meat:
→ Lamb

Fish:
→ Crab
→ Mussels
→ Plaice
→ Salmon
→ Sea trout

*yes, we know, this is technically a veg!

SUMMER

Veg:
→ Aubergine
→ Beetroot
→ Broad beans
→ Broccoli
→ Carrots
→ Chilli
→ Courgettes
→ Fennel
→ French beans
→ Globe artichokes
→ Lettuces
→ Peas
→ Peppers
→ Radishes
→ Rocket

→ Runner beans
→ Salad leaves
→ Samphire
→ Spring onions
→ Sweetcorn
→ Tomatoes

Fruit:
→ Apricots
→ Blackcurrants
→ Cherries
→ Gooseberries
→ Peaches
→ Raspberries
→ Redcurrants
→ Strawberries

Fish:
→ Mackerel
→ Plaice
→ Prawns
→ Salmon
→ Sardines

All the fresh herbs:
→ Basil
→ Chives
→ Coriander
→ Dill
→ Mint
→ Parsley
→ Sage
→ Tarragon

AUTUMN

Veg:
→ Beetroot
→ Broccoli
→ Butternut squash
→ Carrots
→ Cavolo nero
→ Cauliflower
→ Celeriac
→ Celery
→ Chard
→ Chicory
→ Courgettes
→ Fennel
→ Globe artichokes

→ Kale
→ Kohlrabi
→ Leeks
→ Parsnips
→ Pumpkins
→ Radishes
→ Runner beans
→ Sweetcorn
→ Swede
→ Turnips
→ Wild mushrooms

Fruit:
→ Apples
→ Blackberries

→ Damsons
→ Figs
→ Grapes
→ Plums
→ Raspberries

Meat:
→ Beef

Fish:
→ Cod
→ Coley
→ Haddock
→ Plaice
→ Sea bass

WINTER

Veg:
→ Beetroot
→ Brussels sprouts
→ Cabbage
→ Cauliflower
→ Celeriac
→ Chard
→ Chicory
→ Celery
→ Jerusalem artichokes
→ Kale
→ Kohlrabi
→ Leeks
→ Purple sprouting broccoli
→ Parsnips
→ Potatoes
→ Pumpkin
→ Salsify
→ Swede
→ Turnips

Fruit:
→ Apples
→ Blood oranges
→ Cranberries
→ Oranges
→ Pears
→ Pomegranate
→ Rhubarb
→ Satsumas
→ Tangerines

Meat:
→ Duck
→ Venison
→ Turkey

Fish:
→ Clams
→ Haddock
→ Hake
→ Mackerel
→ Mussels

SEASONAL FEASTS

MOB, now we know what's in season when, it's time to find out what to do with some of our favourite seasonal ingredients.

SPRING

Asparagus
The Queen of Spring. When asparagus is in season eat as much of it as you can.

1. To prep, snap off the woody ends with your fingers. There will be a natural place where the asparagus breaks. Keep the ends and boil with potatoes, herbs and stock for a soup.
2. Whack asparagus spears under a hot grill with a little olive oil and seasoning. Grill for 8–10 minutes until tender, serve with poached eggs.
3. Or, chop the spears into small pieces. Fry some pancetta until crisp, add the asparagus, some chopped garlic and lemon zest. Cook for 2–3 minutes until tender, then dollop in some crème fraîche. Toss with cooked spaghetti and a splash of pasta water.

Radishes
Crisp, crunchy, fiery and pink. Freshen up your spring with the humble radish.

1. Wash well. Keep whole with leaves and stems attached – great dunked in hummus.
2. Or, cut in half, drizzle with oil and season with salt and pepper. Roast at 160°C fan (180°C/350°F/Gas Mark 4) for 20 minutes. Sprinkle with za'atar and serve.

Spring greens
More robust than spinach but less crunchy than kale, spring greens are a firm MOB favourite.

1. To prep, trim the base then thinly slice.
2. Drizzle olive oil into a frying pan over a high heat. Add sliced garlic and cook for 30 seconds, then add the greens with a splash of water. Fry until wilted, then season with salt, pepper and lemon juice to taste.
3. Or, add a handful of shredded spring greens through a curry or dahl just before serving.

SUMMER

Beetroot

Earthy and sweet, cook a bunch of beetroot at a time and keep cold in the fridge for the rest of the week.

1. Cut away the leaves. Wash and keep to use in stir-fries.
2. Drop unpeeled beetroot into a pan of boiling water. Cook for 45 minutes– 1 hour until a cutlery knife can be inserted into the middle easily. Drain. Once cooked they will be easy to peel.
3. Or, drizzle the raw beetroot with oil. Roast, skin-on, at 180°C fan (200°C/400°F/Gas Mark 6) for 45 minutes until tender. Cool, then peel, season with salt, pepper and a pinch of ground cumin.

Runner beans

The Daddy of green beans – these are not bang on trend but they've been wrongly forgotten.

1. To prep, top and tail the beans, then slice into thin horizontal slices. Keep the ends for making stock.
2. Drizzle some oil into a frying pan over a high heat. Chuck in the sliced beans and fry for 2 minutes. Pour in 2 beaten eggs, dollop in some pesto, then fold together for a quick summer omelette. It's a Freshy.

Fennel

Don't be put off by the aniseed flavour. Done right, fennel is a worldie of an ingredient.

1. To prep, pick off the green fennel fronds and save for garnish. If eating raw, slice the fennel as thinly as you can into long strips.
2. Toss thinly sliced raw fennel with lots of olive oil, lemon juice, roughly chopped tomatoes and parsley. Season with salt, pepper, chilli flakes and a pinch of sumac.
3. Or, cut the fennel into quarters lengthways. Drizzle a glug of olive oil into a pan. Add the fennel, fry over a medium heat until caramelized and soft. Add to salads or use as a side.

AUTUMN

Pumpkin

MOB – Roast pumpkin is absolutely sublime. Rich. Sweet. Sticky.

1. To prep, cut the pumpkin in half. Remove the seeds and the stringy pith and place in a bowl. Wash and dry the seeds on some kitchen roll, then toss them in your favourite spices and oil. Roast on a baking sheet at 160°C fan (180°C/350°F/Gas Mark 4) for 15 minutes until crunchy.
2. Season pumpkin halves and place cut-side down on a tray. Roast at 180°C fan (200°C/400°F/Gas Mark 6) for 1 hour until completely soft. Serve whole or use the flesh in a soup or as a creamy non-dairy pasta sauce.

Chard

Chard comes in the most beautiful colours and has a rich, irony flavour.

1. To prep, cut the stalks away from the leaves. Thinly slice the stalks and keep the leaves whole. The stalks take longer than the leaves to cook.
2. Put a sheet of puff pastry into the oven at 180°C fan (200°C/400°F/Gas Mark 6) for 20 minutes. Melt a knob of butter in a pan, add the chard stalks and fry for 5 minutes, then crush in some garlic. Cook for a further minute, then throw in the chard leaves. Once wilted, season with lemon zest, salt and pepper. Pile on top of the cooked puff and drizzle over some tahini yogurt.

Celeriac

This knobbly veg looks ugly, BUT it tastes sweet, nutty and delicious.

1. To prep, trim the muddy base from the celeriac and peel the rest. The skin is quite thick, so using a knife may be easiest. Wash the peelings and keep them to make crisps or add to stock (see pages 97–98).
2. Cut the celeriac into bite-sized pieces. Drizzle with oil, salt and pepper. Roast at 180°C fan (200°C/400°F/Gas Mark 6) for 35–40 minutes until soft.
3. Mix the roasted celeriac with some cooked Puy lentils and wilted kale. Drizzle over a splash of vinegar – balsamic would be banging. Top with soft-boiled eggs and some toasted hazelnuts to finish.

WINTER

Celery

Celery gets a bad rap as a health food, but don't shame him, he's so versatile.

1. To prep, wash the stalks well. Pick off the leaves and save them to use in salads or as a garnish.
2. Chop the celery stalks finely along with carrot and onion. Slow-cook in loads of olive oil to make the base for a rich tomato sauce (sofrito).
3. Or, cut the celery into finger-length pieces, then fry in a knob of butter with garlic and thyme until starting to brown. Pour over enough stock to cover, whack on a lid and simmer for 20 minutes until soft. Serve as a side.

Swede

Super cheap and filling with a slight peppery flavour. Swede is a veg you need to get on board with.

1. Wash and peel. Keep the peelings for crisps or veg stock (see page 98).
2. Cut into chunks, boil until completely soft, then mash with a knob of butter.
3. Or, roast swede chunks at 180°C fan (200°C/400°F/Gas Mark 6) with olive oil, chilli flakes and ground cumin for 35 minutes. Drizzle over some maple syrup, then return to the oven for 5 minutes until caramelized.

Chicory

On its own chicory can be bitter, but dress it right and it is out of this world.

1. Separate the outer leaves and thinly slice the core.
2. Dress both with a little mustard, balsamic, lemon juice and olive oil. For an epic salad, mix with chopped apple, toasted walnuts and feta.
3. Or, cut the whole chicory into quarters lengthways. Season, drizzle with oil and a little vinegar. Roast at 180°C fan (200°C/400°F/Gas Mark 6) for 10 minutes until soft, then pour over double cream and grate over some Parmesan. Slide under the grill for 5 minutes until bubbling and golden. Delicious.

DIET

There's been a lot of noise lately about what the best diet for the planet really is, be it veganism, flexitarianism or the recently proposed planetary health diet. There are some strong opinions out there and it can sometimes be quite confusing.

We believe everyone should have the freedom to eat what they want and not be judged on their food choices. This chapter is about giving you the information you need to make informed decisions and get on with your life.

VEGAN VS. PLANT-BASED

You've probably heard the words 'plant-based' and 'vegan' being thrown about a lot recently, but what do they actually mean and are they really the same thing?

The simple answer is 'no'. Veganism is a way of life, while plant-based is focused on what you eat.

Vegans try to avoid using any product where an animal has been involved in the process of making it, from cosmetics that use animal by-products like beeswax, to clothing and shoes made with leather, wool or silk.

A plant-based diet is one that is focused around plants, with few-to-no animal products being consumed. Whilst this means that plant-based diets aren't necessarily vegan or vegetarian, the term 'plant-based' is often used to label items that are in fact vegan, be it on restaurant menus or on packaging.

The key thing to remember is that plant-based foods are vegan BUT being on a plant-based diet isn't.

Vegan pros:
→ Brilliant for those conscious of **animal welfare**.
→ Encourages you to eat a variety of healthy foods with **lower saturated fats**.
→ Better for the **planet** (see next page).

Vegan cons:
→ Requires **research and planning** when buying and eating out.
→ Meat alternatives can be **expensive** and highly processed.
→ Can lead to **deficiency** in some proteins, vitamins and minerals.

FLEXITARIAN

A good halfway house. (This is similar to the pescetarian diet, where you cut out meat but carry on eating fish.)

As the name suggests MOB, this way of eating is all about flexibility. It means you have a mostly vegetarian diet and will occasionally eat meat or fish too. It's a great, well-balanced and non-restrictive approach.

Being flexitarian is becoming an increasingly popular choice for people who care about the environment but don't want to become full-on vegan. It doesn't mean just having one less chicken burger a week, a flexitarian diet recommends eating 203g/7 oz chicken per week – that's about the size of 1½ chicken breasts.[19]

Flexitarian pros:
→ **Supporting** all types of **farmers**.
→ Socially **inclusive**.
→ **Beneficial** impact on the planet.

Flexitarian cons:
→ Relaxed rules require a lot of **willpower** to stick to.
→ Can be **tempting** to give yourself more **meat and fish** than the recommended portion amount.
→ Requires **planning** to make meals vegetarian if you are not.

PYRAMID OF IMPACT

So now you know how these different diets work, it's time to look at which of these dietary changes will have the biggest impact on our planet. The diagram below demonstrates just how much of an influence each diet has on reducing our food-related emissions.[20]

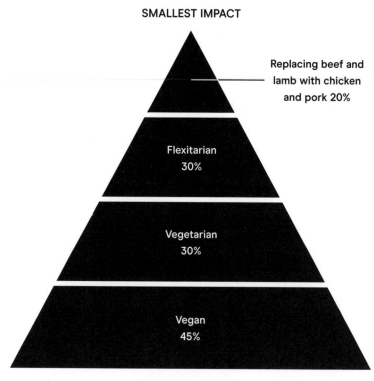

SMALLEST IMPACT

Replacing beef and lamb with chicken and pork 20%

Flexitarian
30%

Vegetarian
30%

Vegan
45%

GREATEST IMPACT

WHAT'S THE BEEF?

In 2014, the groundbreaking documentary *Cowspiracy* exposed the destructive impact of animal agriculture on the planet. Since then, the film has come under fire for exaggerating and cherry picking the 'truth'. Still, cows and their environmental impact continue to stay in the spotlight.

So why cows?
Put simply, when you are the biggest beast you require the most energy. You need more food, more water and more land to grow.

Then, as cows digest their food, they naturally go through the motions. The problem is, in doing so, they produce the greenhouse gas methane.

Of course, it's not just cows that produce methane; sheep, goats and even wild giraffes and camels have similar digestive systems and produce their share of methane too. But cows are the primary offenders.

What's the impact?
Globally, cows are the main source of livestock greenhouse gas emissions – around 65–77%.[21]

The Intergovernmental Panel on Climate Change (IPCC), the official authority on climate change chat, has identified beef as the single food with the greatest impact on the environment.[22] For example, in the US 4% of all food sold is beef, which alone contributes to a whopping 36% of the USA's food-related emissions.[23]

So, should we just stop eating beef and dairy then?
It's not the cows' fault. It's the way we farm them. Intensive farming, where huge numbers of cows are kept and fattened quickly with little-to-no room to roam = bad news. Regenerative farming, where cows are used to enhance the ecosystem by using their waste for fertilizer and allowing them to roam = good news.

HOW CAN WE HELP?

At MOB Kitchen we are striving to eat less meat and ensuring that when we do, we're buying the best quality we can afford. This means we're eating lots more fruit and veg. If this is new to you, here are some hacks.

1. Start at breakfast

Chances are, unless you are rolling in it, you're not eating meat for breakfast every day. To continue this, can you challenge yourself to a whole month with no bacon sarnies?

2. Comfort food swap

What's your favourite comfort food? If it's spag bol, then why not have a go at our lentil and walnut version instead? Or maybe it's a burger you are after? Whatever it is, we've got all your needs covered at **www.mobkitchen.com.**

3. Choose a veggie recipe

Just one veggie recipe is all you need to get started. Pick a recipe that interests you and have a go. If you like it and it becomes part of your weekly rotation, then that's one less meat meal you are eating a week. Pick up our *MOB VEGGIE* cookbook for inspo.

4. It's all about flavour

The key to banging vegetarian food is to maximize flavour. No one likes an over-boiled carrot. This means roasting vegetables in spices and making punchy sauces and dressings. A veggie curry is always a good place to start.

5. Count your money

This is some real motivation MOB. See how much money you are saving each week as you buy less meat.

6. Meatless Mondays

Once you've got the hang of it, try doing a couple more days every week.

PROTEIN

One thing we all keep hearing is less meat = less protein. But it's actually not as clear cut as you would think. The infographic (right) shows you the average amount of protein you receive per 100g of each ingredient.[24]

STEAK 25.5g
EDAMAME BEANS 12g
POTATOES 2g **CHICKEN**
WHOLE COW'S MILK 3.5g **THIGHS 28g**
PARMESAN CHEESE 32g
MOZZARELLA CHEESE 17g **EGGS 14g**
PORK LOIN 27.5g
LAMB MINCE 21.5g
PEANUT **RED LENTILS 8g**
BUTTER 28g
SALMON 22g
QUINOA 5g
CHEDDAR **TOFU 12.5g**
CHEESE 25.5g
CHICKEN **CHICKPEAS 6g**
BROCCOLI 4g
BREAST 30g

SUSTAINABLE FOOD CHOICES

Now we've covered diets, it's time to address sustainable food choices. This was one of the areas we found trickiest – working out how to buy better.

We had lots of questions, mainly:
→ What is really going to make the most difference?
→ What's the deal with food miles?
→ Why is one brand of chocolate biscuits okay, but not the other?
→ If I'm going to eat less meat, but then eat more soy, is that even better for me or the planet?
→ Organic is more expensive, but why?

We've sifted through all the noise, so there's no need for you to be confused as well. Here are the key things you need to know. You're welcome.

FOOD MILES

We live in a world where food has become globalized. We have the choice of buying any ingredient all year round. Imagine walking into the supermarket and not being able to buy whatever you want, whenever you want.

So, what are the implications of eating food grown on the other side of the world, and what exactly are food miles?
Food miles refer to the distance food has travelled from where it was produced to when it reaches your plate.

A typical basket containing 26 imported goods may have travelled a distance equal to 6 times around the equator.[25]

Is this a problem?
Yes and no. Food miles alone account for 11% of global food greenhouse gas emissions.[26] But not everything grown abroad has a worse environmental impact. Take a tomato sold in the UK in winter. It takes less energy for that tomato to be grown in an open field and shipped from Spain than it does to grow it in the UK in a heated greenhouse.[27]

However, in general, buying food from around the world means that it's of poorer quality. Fruit and veg are often picked underripe so that they don't go off on the journey.

The prices of these global foods are also kept low by supermarket competition. If you think about it, if we are paying less than £2 for some mangetout that's been transported across the world, then how much is the farmer who grew it getting paid for it?

But how can you tell where your food is from?
Packaged fresh ingredients will have a country of origin on the label. If picking up loose fruit and veg, check the label on the crate itself.

So, remember to eat seasonally and support our local farmers. To find out more about the food miles of specific ingredients, head to the website **www.foodmiles.com**.

PALM OIL

What is it?

Palm oil is an edible vegetable oil that comes from the fruit of oil palm trees. Native to West Africa, it was brought over to South East Asia and has become, due to its productiveness, the largest oil-producing crop on the planet.

What is palm oil used for?

It's the secret ingredient in SO many everyday things. Pizzas, toothpaste, instant noodles, ice cream, processed bread and chocolate – in fact, it is in up to 50% of all packaged products we buy in the supermarket.[28]

What's the problem with it?

Palm oil is one of the leading causes of deforestation along with animal agriculture. This deforestation is destroying the homes of some of the world's most endangered species, such as the orangutan, Sumatran rhino and pygmy elephant.[29]

So, should we stop eating it?

Unfortunately, there isn't a simple answer. While in the West palm oil is used as a cheap bulker, in West Africa it has provided a vital source of food and fibre to people for over 4,000 years.[30]

Millions of smallholder farmers in Southeast Asia rely on palm oil for their livelihoods. Boycotting palm oil altogether would mean putting a lot of vulnerable people out of work.

Could we just use another crop?

If we were to replace palm oil with another oil crop, we would need to clear an area the size of Spain in order to produce the same amount of oil.

So, what's the solution?

1. Buy sustainable palm oil. Look out for the Roundtable on Sustainable Palm Oil (RSPO) stamp, which is globally recognized.
2. Make smart swaps. If your usual brand of chocolate biscuits contains non-sustainable palm oil, then look for an all-butter variety instead.

SOY

The soy or soya bean is a type of legume native to East Asia. Argentina, Brazil and the US are now responsible for 80% of the world's soy production.[31]

There are many varieties of soy. For instance, did you know that edamame are in fact green soya beans?
Soya is extremely versatile and is the base of some of these beautiful ingredients:

→ Soy sauce
→ Soya milk
→ Tofu
→ Miso
→ Tempeh

Soy is a great source of plant-based protein, B vitamins and minerals such as iron and zinc. As a seasoning in soy sauce it adds bags of flavour and in its more solid form, as tofu, it can be used as a meat substitute.

So why is it controversial?
Back in the '90s, scientists found that soy contains isoflavones (we didn't know what these were either.)

Isoflavones are a chemical compound with similar properties to oestrogen, the female sex hormone. When people discovered this, soy was initially shunned. However, since then further research has been done to allay health fears and soy is now regarded by some as having excellent health benefits including reducing the risk of breast cancer and improving heart health by lowering cholesterol levels.[32]

What about the environment?
At the start of this century, soy was one of the leading causes of deforestation in the Amazon, threatening one of the planet's most vital ecosystems.[33]

This changed in 2006 when leading world companies signed the Amazon Soy Moratorium to limit deforestation by restricting soy production to existing agricultural land.[34]

Soy is still grown intensively though. As our appetite for meat increases, so too does the need for soy – 75% of the world's soy is actually grown in order to feed livestock.[35]

So, what can we do?
As bizarre as it may sound MOB, cutting down on meat will have the biggest impact on reducing industrialized soy production. There's your answer.

MILKING IT

Are you an oat milk flat white kind of person?
Or do you prefer a cappuccino made with whole milk?

We've got the ultimate breakdown on which milk has the biggest
environmental impact[36]... best to get more into your oat milk then.

Cow	3.2	9.0	628
Almond	0.7	0.5	371
Oat	0.9	0.8	48
Soy	1	0.5	28

WILD FISHING

There are two main types of fishing: wild and farmed, and over the next four pages we've laid out the pros and cons of each method.

THE BAD NEWS

→ **Overfishing:** According to the United Nations, 90% of the world's fish are overexploited. Rising demands threaten to wipe out entire species completely.[37]

→ **Habitat destruction:** Types of fishing known as dredging and bottom trawling use weighted nets to pick seafood and some fish from the seabed, wiping everything else out at the same time. This is just about as bad as it gets.

→ In tropical regions, illegal **blast fishing** is still happening. Fishermen throw lit sticks of dynamite into the water; catching shed loads of fish and destroying coral reefs in the process.

→ **Bycatch:** Bycatch are unwanted fish and other marine creatures that get caught in fishing nets.

→ Every year, 300,000 marine mammals, 160,000 albatross and 3 million sharks are lost to bycatch. **Every. Single. Year.**[38]

THE GOOD NEWS

→ **Global laws:** Across the world, governments are taking action. Laws are being made to limit the amount of fish that fishermen can catch and what fish they can catch each season.[3]

→ **Sustainable Fisheries:** In 2020 a new bill was passed in the UK to ensure that fish stocks and the marine environment will be better protected for future generations.

→ **Eco-friendly fishing methods:** Fishermen are adopting different fishing methods to ensure minimal environmental impact, such as using massive fishing rods. Fish caught in this way are better for the environment and better quality.[40]

→ **Old tech:** In many indigenous cultures, people have fished sustainably for thousands of years. Today's sustainable fishing practices reflect some lessons learned from these cultures, such as the use of cast-net and spearfishing.

→ **New tech:** From acoustic pingers which emit sounds that marine animals hate, stopping them from swimming into nets – to putting grates in shrimp nets in Mexico that turtles are unable to swim through, new technology is a game changer that can help prevent bycatch.[41+]

FARMED FISHING

THE BAD NEWS

→ **Fish feed:** More fish being farmed = a lot of mouths to feed. Farming carnivorous species such as salmon and prawns requires enormous amounts of fish feed, often made from wild fish – putting more pressure on the ocean.

→ **Disease and pollution:** When farmed inhumanely, as many fish as possible are crammed into small enclosures. In these conditions diseases and parasites spread. A lot of fish in a confined space = a lot of waste. When not properly managed, this can pollute the water.

→ **Habitat destruction:** Sensitive natural habitats converted for fish farming are bad news for the environment. Prawn farms are often set up in areas that were once mangroves.

It is massively important to preserve these areas MOB, as mangroves are one of the most effective ecosystems at storing carbon.[42]

THE GOOD NEWS

→ **New technology:** Farmers can now minimize waste with no disturbance to the fish. In some places land-based growing tanks are being built to save water, using recirculating water systems.[43]

→ **Healthcare:** Endless expert research is being done into maintaining the health of farmed fish to stop the spread of disease.[44]

→ **Learning lessons:** Fish farmers are now learning from their mistakes and progressing.

→ Some feed producers are experimenting with using insects rather than wild fish for food.[45]

What can we do MOB?

Support marine conservation charities. Our favourites are the Marine Conservation Society (MCS)[46] and MBA Seafood Watch.[47] Both websites have all the info you need, as well as regularly updated guidelines telling you what fish you should and shouldn't eat.

Buy responsibly sourced. This is the biggie, MOB. Check the packets for a Marine Stewardship Council (MSC)[48] or Aquaculture Stewardship Council (ASC)[49] tick when buying fish and seafood.

BEHIND THE LABELS

Food labels are often difficult to understand. So, MOB, consider this the need-to-know info.

ORGANIC:

Soil Association and Organic means that the food has been produced with:
→ No genetically modified ingredients
→ Fewer pesticides
→ No artificial colours or preservatives
→ No routine use of antibiotics

Organic = naturally produced.

To get an organic certification farms need to pass an intense inspection at least once a year.[50]

This means that organic food is:
→ Always free range
→ Produced with higher levels of animal welfare
→ More environmentally sustainable with better land management and protection of wildlife
→ Traceable – you can find out where and who grew your food

And, importantly...why is it often more expensive?
Without the use of manufactured fertilizers the food is harder to grow.

Look out for own-brand supermarket ranges. Try swapping one of your basics such as milk, butter or carrots to organic.

FREE RANGE:

Free range means the animals are allowed to roam freely outside for at least part of the day.[51]

The amount of space and time the animals have outside can vary dramatically. Try to look for a free range symbol to help improve animal welfare.

FAIRTRADE:

Fairtrade protects poorer, vulnerable workers and farmers from discrimination. It sets standards that they and their companies have to follow.[52]

Farmers and workers must:
→ Maintain their workers' rights
→ Protect the local environment

Companies must:
→ Maintain decent working conditions
→ Pay better – the Fairtrade minimum wage
→ Pay a Fairtrade Premium

What is a Fairtrade premium?
Companies pay an additional sum of money to their farmers and workers that goes into a communal fund for them to use as they wish.

Fairtrade also helps tackle specific issues facing a community like investing in a woman's leadership school in West Africa[53] – epic.

In short, Fairtrade is a winner. Look out for Fairtrade chocolate, coffee and tea.

RSPCA ASSURED:

Royal Society for the Prevention of Cruelty to Animals (RSPCA) assured is a scheme dedicated to improving the welfare of farm animals.[54]

To get this label farmers must:
→ Monitor the health and welfare of their animals
→ Give animals bedding
→ Provide the animals with greater space
→ Slaughter animals humanely

These rules apply to both indoor and outdoor reared animals.

Make sure you look out for meat with this great label.

RED TRACTOR:

Red Tractor is a scheme that maintains quality standards for food safety and hygiene within the UK.[55]

To get a Red Tractor label farmers must:
→ Maintain animal welfare
→ Limit their use of antibiotics
→ Use humane slaughter guidelines

Red Tractor provides a benchmark for the farming industry, setting the minimum standard for how livestock should be raised.

BRITISH LION:

The British Lion is a quality mark for eggs. They are produced to such high standards that they are approved by the Food Standards Agency to be eaten runny or even raw.[56] Most eggs in the UK are British Lion, but not all, so look for the iconic Lion printed on the shell.

Recently, British Lion introduced a higher standard for barn eggs, to help supermarkets move to cage-free eggs by 2025.

We love the British Lion egg.

SUSTAINABLE SWAPS

MOB, sometimes simple changes like swapping out ingredients is enough to make a little difference to our environmental impact. Here are our top three swaps:

BEEF
⟹
CHICKEN

Beef requires on average 28 x more land, 11 x more water and 6 x more fertilizer than chicken.[57] Choosing chicken over beef will cut your carbon footprint in half! Go on. Sack off the beef and make a difference.

Farming prawns requires a ton of resources, whereas farming mussels and clams has some of the lowest carbon emissions. In fact, these small shelled beauties can actually improve the water quality of their environment.[58] And they are, without doubt, more delicious. It's a no-brainer.

PRAWNS
⟹
MUSSELS/ CLAMS

ALMONDS
⟹
HAZELNUTS

A single almond takes 5 litres (8 pints) of water to grow![59] Hazelnuts require very little water and minimal upkeep. These clever nuts are even drought resistant, can survive in harsh conditions and even help with soil erosion.[60]

SHOPPING AND ORGANISATION

Now we're going to put this new information into practice to make your kitchen more sustainable. Let's get on it MOB.

First things first and that's getting organized. All it needs is one big session, get some tunes on and rope in your housemates to help.

Becoming more sustainable begins at the supermarket. We're here to keep you on the straight and narrow with shopping advice, so you only buy what you really need.

SORT YOUR CUPBOARDS

1. **Get everything out.** It's going to create a massive mess and look worse than before, but you've got to get it all out. The old spices. The weird packets of noodles. All of it.

2. **Give those shelves the clean of all cleans**, so they're ready for what's to come.

3. **Condense.** Are there four open bags of rice with only a tiny bit left in each? Pour them all into the same bag, or better yet empty them into a reusable container. Each bag of rice may belong to a different housemate, but having tiny bits of this and tiny bits of that is a surefire way for food not to get eaten and to get chucked in the bin. Group mentality is key.

4. **Communal space.** On that last point, ask your housemates if they would be happy sharing dried basics such as pasta and rice. If the answer is yes, allocate one person to be in charge of replenishing each ingredient. This works out cheaper and is more efficient at reducing waste.

5. **Sort out the spices.** Pool all your spices together and work out what you've got tons of and what you need. Four smoked paprikas? Well you know you're not buying that for a while. Put all the spices back in one place so that everyone can use them when they cook.

6. **Personal space.** Clearly some food items aren't for sharing. Work out where the best place in your cupboard is for communal things and then divide the rest of the space up between you. It's going to be a smaller space than before but consider it a challenge to buy less.

7. **Putting things back.** Put the things you've got a lot of at the front of the cupboard so that they are the easiest to see. This will encourage people to cook with what you already have.

8. **Challenge:** As a house, see if you can use up half of all your communal dried ingredients before you buy any more. Again – less spending, less waste. You know it makes sense.

THE SHARED FRIDGE

Ever looked inside the fridge and seen a bag of spinach slowly go off until it has to be thrown away? You would have eaten it, only it was your housemate's. It's time to change that – here's how to get the most out of your fridge.

1. **Everything out.** As with the cupboards, start by seeing what you've got.

2. **Clean.** It's EVEN MORE important here because a clean fridge stops smells from spreading and can keep ingredients fresher for longer.

3. **Is it off?** This book is all about cutting down on waste, but before you can start that, it's time to get rid of anything that has clearly had its day. We aren't talking a bit of feta that has just gone past its sell-by date (more on that later – see pages 82–83) – to throw something away it needs to be clearly passed the point of no return.

4. **Condiments.** If your housemates are up for it, can you make at least part of your fridge communal? First up being an area for condiments. If you have two of the same thing, condense and then rearrange the condiments in your fridge, putting anything with only a small amount left at the front so you know to use it up first.

5. **'To eat' shelf.** This is the most important step to stop food waste MOB. Allocate a space in your fridge where all of you will put ingredients that need to be used up. This is the place where you put that bit of leftover yogurt when you are going away for the weekend. If food is placed here people know to eat it first.

6. **Share the essentials.** Are there essentials that everyone in the house uses? If so, allocate one person to be in charge of replenishing the house milk, for example. That way you will never end up with multiples of the same thing taking up room in the fridge.

SUPERMARKET SURVIVAL

Now you've organized the kitchen MOB, it's time to go shopping. Smashing the shopping and buying only what you need is key to reducing food waste. These are our top tactics for holding it down in the store:

1. **Never shop hungry.** We've all been there, starving for some scran in the supermarket and shoving everything into your basket. It's a fatal error; have a snack before you go.

2. **Check what's in your kitchen.** What have you already got at home? We don't just mean fresh ingredients. Look in your cupboards before you head out and make a note of what you can cook with. You may already have most of what you need.

3. **Plan.** Take 10 minutes to roughly plan what you are going to cook for the next few days. Be realistic about when you are actually going to be eating in.

If you know you are going to be home late a few nights, think of a meal that you can batch cook to be reheated quickly when you want it.

4. **Make a note.** Write a list of everything you need for your meals, then walk around the kitchen and delete anything you already have. Check your phone in the supermarket to remind yourself of the list and stop you going off track.

5. **Avoid multi-buys.** Do you really need four of that? Just because it looks cheaper doesn't mean it's the best option if it isn't something you need. Stay away from buying deals for the sake of deals. Buying too much means you are more likely to waste it later.

6. **Take your reusables** – don't forget your bags!

MEAL PLAN

Start using a meal planner and become ruthlessly efficient at batch cooking and reusing those leftovers.

	BREAKFAST	LUNCH	DINNER
MON			
TUE			
WED			
THUR			
FRI			
SAT			
SUN			

PACK IT IN

Now it's time to turn our attention to packaging. Later on, we've got a whole chapter on plastic (see pages 108–115) – but here are our top ways to buy less plastic, because there is nothing more depressing than walking out the store with armfuls of packaging and only a few ingredients.

1. **Go big**. Roping your housemates in to sharing communal ingredients means you can buy in bulk for those long-life dried ingredients you use all the time. Buying in bulk = smaller packaging to food ratio.

2. **Loose fruit and veg.** Fruit and veg have their own natural packaging – their skin. Wherever possible avoid buying it in a plastic bag.

3. **Avoid multi-packs.** Let's take a four-can pack of chopped tomatoes, for example. The tins themselves = recyclable. That pointless piece of plastic wrapped around the outside of them = non-recyclable.

4. **Say NO to ready meals.** Even if you are unbelievably tired and can't be bothered to cook, a fried egg on toast will take you less than 5 minutes. Step away from the plastic aisle.

5. **Make plastic swaps.** Head to page 112 for our best ones.

6. **Buy from the counter.** There often tends to be cheaper supermarket deals on counter items, like cheese for example. You only buy what you need and it comes wrapped in paper. PLUS, if you bring a container with you, they'll put the food into that.

7. **Take your reusables** – don't forget your bags.

MONEY TALKS

Getting organized, sharing food, planning your shopping and buying less isn't just better for the environment folks – it will also save you money. Spend the savings on a meal out with the MOB.

COOKING

With all this said, one of the easiest and most effective ways of being more sustainable in the kitchen is cooking.

Becoming a more confident cook will not only save you money – less takeaways – but will reduce your packaging consumption – fewer ready meals – and help reduce your food waste. Plus, there is nothing more satisfying than looking in your fridge and making a meal out of 'nothing'.

It may require a bit of trial and error. Your first odds-and-ends soup may taste a bit dodgy, but you'll get there if you give it a go. You can't be called a good cook until you are able to create a dish out of nothing.

This is our guide for getting the most out of your ingredients. Let's get to it.

INTERCHANGEABLE

Have you ever looked at a recipe that calls for soured cream when you only have yogurt in the fridge? Did you go out and buy the soured cream worried that the meal wouldn't taste the same without it?

At MOB Kitchen we love trying new ingredients, but making smart swaps based on what you've already got in the fridge is a great way of improving your cooking confidence and learning how flavours work. Most of the time small ingredient switches won't make a big difference to the end result.

Learning to interchange ingredients is the first step to becoming an intuitive cook, and this list shows you how.

Butternut squash ↔ sweet potato

Broccoli ↔ cauliflower

Peas ↔ sweetcorn

Smoked bacon ↔ pancetta

Pecorino ↔ Parmesan

Gruyère ↔ Comté

Mangetout ↔ sugar snap peas

Crème fraîche ↔ soured cream
or yogurt

Chopped tomatoes ↔ passata

Mixed spice ↔ allspice + pinch of
ground coriander

Saffron ↔ turmeric (colour) +
pinch of smoked paprika (flavour)

Balsamic vinegar ↔ red wine vinegar +
pinch of sugar/drop of honey

INGREDIENTS

Buttermilk ↔ whole milk + squeeze of lemon juice (leave for 15 minutes until the milk starts to curdle)

Creamed coconut ↔ coconut milk (dilute the creamed coconut with boiling water)

Maple syrup ↔ honey

Self-raising flour ↔ plain flour + baking powder (½ tsp baking powder for every 100g plain flour)

Curry powder ↔ garam masala + pinch turmeric + chilli powder

Cayenne pepper ↔ chilli powder or smoked paprika + pinch of chilli flakes

Ras el hanout ↔ mixed spice + big pinch of ground cumin + smoked paprika

Harissa ↔ chilli paste + 1 tsp tomato purée + 1 finely chopped garlic clove

Lemons ↔ limes (just use a little less as lime juice is usually tarter)

Green chilli ↔ red chilli

Basmati rice ↔ long-grain rice

1 red onion ↔ ½ onion (red onion is sweeter, so you need less normal onion)

Cannellini beans ↔ butter beans or haricot beans

Swede ↔ turnip

Kidney beans ↔ black-eyed beans or pinto beans

Spaghetti ↔ linguine or tagliatelle

BREAKFAST HACKS

A decent breakfast always gets you going for the day, but when you're running out the door in the morning it can slip your mind. Sooner or later when hunger hits you're bound to buy something. Be it a yogurt pot, sandwich or cereal bar; whatever it is, it will come in packaging.

This packaging, with a little bit of forward planning, can easily be avoided. All that's required here is a bit of effort the night before. Here are some of our top breakfast hacks.

OVERNIGHT OATS:

1. Half-fill a clean jar with oats, then cover completely with water or milk of your choice.
2. Add flavour: handful of frozen berries/ 1 grated apple + dried fruit/1 sliced banana + pinch of cinnamon.
3. Add toppings: spoonful of yogurt, honey/maple syrup or nut butter.
4. Screw on the lid and leave overnight in the fridge with a teaspoon resting on top of the jar so you don't forget it.

By the morning, the oats will have soaked up the liquid and become soft and creamy. So good.

BOILED EGGS:

Boil 6 medium eggs for 6½ minutes for a jammy, runny yolk or 8 minutes for nicely hard-boiled.

Once cooked, cool completely in cold water, then put in the fridge. In their shells the eggs will last for 3 days – perfect for grab and go.

BOILED EGG BAGEL:

The night before, grab a bagel, butter it and drizzle over some hot sauce. Put into your Tupperware with two boiled eggs in their shells and a handful of spinach. Construct the bagel wherever you are heading. No more soggy sandwiches.

FRITTATA:

A MOB classic. Head to our website for all our frittata gems. They all make enough for four, so smash one out on a Sunday for your dinner and then keep the other portions to have for breakfast Mon-Wed. Sorted.

→

POWER SMOOTHIE:

Blitz 1 banana, 1 tbsp nut butter, 1 large handful of spinach and 200ml water or milk of your choice. Add some oats if you want it to be more filling.

MOB'S 10/10 GRANOLA:

This is a game changer. It will save you money and make morning munch easier and tastier:

1. Mix 1 box of oats with some seeds, dried fruit and/or nuts. Add 2 tbsp each of ground cinnamon and ginger. Pour in 100ml coconut or olive oil and around the same again of honey or maple syrup. Add a big pinch of salt. Mix everything together.
2. Bake in an oven at 140°C fan (160°C/325°F/Gas Mark 3) for 30–40 minutes until deep golden – mix every 10 minutes so that the edges don't burn. Keep in an airtight container for up to 2 months.

MOB SNACKS

We all need a snack from time to time. Making your own is an easy way to save on plastic – learn these and you're good.

HUMMUS:

Shop-bought hummus is no match for the real thing. Empty a 400g (14oz) tin of chickpeas along with half of their liquid (keep the rest, see page 95) into a food processor. Add 2 tbsp tahini, a splash of olive oil and some seasoning. Blitz until super smooth and creamy. This hummus keeps for up to 5 days in an airtight container in the fridge.

Flavour ideas: Add ½ tsp smoked paprika or ground cumin, blitz in 1 small garlic clove and/or squeeze in some lemon juice.

PEANUT BUTTER AND APPLE:

As simple as they come. Put some peanut butter in a Tupperware with a whole apple (uncut, so it doesn't go brown) and bring a cutlery knife for spreading/slicing.

POPCORN:

Heat some oil in your most indestructible large saucepan over a medium–high heat. Tip in a small bowl's worth of popcorn kernels, stir with the oil, then immediately put on a lid. Keep cooking with the lid on until the popcorn stops popping, then turn off the heat and mix with your chosen flavour. Once cool, keep in an airtight container for 3–4 days.

Flavour ideas: Go spicy with 1–2 tsp chilli powder + ½ tsp ground cumin + salt and black pepper. Like it smoky? 2 tsp smoked paprika + salt works a treat. For chocolate lovers, try 1 tbsp cocoa powder + 2 tsp icing sugar + pinch of sea salt. Or go for a classic combo with 1 tbsp icing sugar + 1 tsp ground cinnamon.

Make sure to keep the flavourings dry so the popcorn will last longer.

FRUIT AND NUT CHOCOLATE:

Melt a big chocolate bar of your choice in a heatproof bowl in the microwave in 30-second bursts, stirring in between each one. Once melted, spread it out onto a piece of baking parchment, then top with your choice of dried fruit and nuts. Leave to set, then break into shards. It will last for up to 2 weeks in an airtight container. Delicious.

Sometimes, you might not have the time to make a snack. On these occasions, opt for things wrapped in paper or foil rather than plastic, such as a big bar of chocolate for instance.

When buying snacks, bigger is usually better. Large bags = less packaging to food ratio. Individual snack packs = huge waste of packaging. Cut it out.

SALAD STATION

Key section. You can always build a good salad from random ingredients you have in your kitchen MOB. Sack off the sad lettuce and cucumber salads... Of course, throw your lettuce and cucumber into the mix, but add some spiced chickpeas and a dollop of yogurt to go with it.

It's not about buying a whole lettuce or needing a massive amount of feta – this is cooking for one. It's all about using up the food you've already got. Follow the steps on the right and get creative.

CHOOSE A BASE	CHOOSE A VEG/FRUIT
Cook off a portion's worth of carbs, if needed, before assembling.	Use whatever you have lying around. If your greens are looking particularly sad, fry them off.
Rice	Green leaves (spinach/watercress/rocket)
Drained and rinsed tins (chickpeas/beans)	Standard raw veg (tomatoes/peppers/avocado)
Any noodles	Roasted veggies (cauliflower/squash/beetroot)
Grains (quinoa/pearl barley/mixed grains)	Hummus dippers (cucumber/carrot/celery, etc.)
Pasta or couscous	From the freezer (peas/sweetcorn/edamame beans)
Potatoes (roasted/boiled)	Flash-fried greens (broccoli/green beans/asparagus, etc.)
	Robust leaves (baby gem/chicory/iceberg)

CHOOSE A PROTEIN	CHOOSE A CRUNCH	MAKE ONE DRESSING	CHOOSE HOW TO FINISH IT OFF
Add some sustenance. You don't need loads here as protein is a topper rather than the main event.	All salads need to have some good texture MOB.	This is the game changer. Make to your taste. A good rule to follow is 2 parts fat to 1 part acid.	This is the point where you ramp up the flavour.
Eggs (fried/boiled/ poached)	Any toasted seeds	Olive oil + balsamic	Salty bits (capers/anchovies)
Hard cheese (Parmesan/ Cheddar)	Fruit (apple/pear/ pomegranate seeds)	Tahini + water + crushed garlic + lemon	Heat (chilli/harissa/ chipotle paste)
Responsibly sourced fish (tinned tuna/ cooked salmon)	Bread (croûtons/ toasted pitta)	Peanut butter + hot sauce + soy + lime	Fresh herbs (mint/coriander/dill/ parsley/basil/sage)
Vegan (tofu/hummus/nut butter)	Any toasted nuts	Olive oil + orange + mustard + honey	Pickled onions (scrunch sliced red onion with lime juice)
Dairy (yogurt/crème fraîche/soured cream)	Veg (kale/radishes/ shredded cabbage)	Olive oil + lemon + chilli flakes	Deli bits (pesto/sun-dried tomatoes/olives)
Meat (cooked chicken/ cooked chorizo/ ham)		Sesame oil + soy + lime + fresh ginger	
Softer cheeses (feta/goat's/fried halloumi)		Olive oil + red or white wine vinegar + mustard + crushed garlic	

SOUP STATION

Soups are the same as salads, MOB – food waste lifesavers. They are especially handy when the stuff in your fridge has seen better days. Blitz up those leftover bits with some seasoning and spices and you've got a worldy of a dish. It couldn't be easier.

CHOOSE A BASE LAYER	CHOOSE HOW TO BUILD THE FLAVOUR
The start of building a great soup before you add your spices and seasoning.	This is where you determine what path your soup is going to take.
Soup trinity: onion + carrot + celery	Pastes (Thai green or red curry/tomato/miso)
Any type of onion (red/white/leeks/ spring onion)	Heat (harissa/chipotle/ chilli/cayenne)
Garlic and... (ginger/anchovy)	Whole spices (cumin/fennel/ peppercorns/ coriander/ cinnamon, etc.)
Stalks and ends (broccoli stems/ chard and kale/ asparagus ends)	Woody herbs (bay leaves/dried oregano/rosemary/ thyme)
Meat (chorizo/streaky bacon/pancetta)	Ground spices (smoked paprika/ turmeric/curry/ coriander/cumin, etc.)
	Tins (coconut milk/ coconut cream/ chopped tomatoes)

CHOOSE THE MAIN EVENT	CHOOSE A FILLER	CHOOSE HOW TO FINISH IT OFF
This is the bulk veg your soup is going to consist of. Once added, pour in water/stock.	If you are not blitzing the soup, you may want to throw in a carb to fill you up.	The final flavour-packing moment.
Roots (sweet potato/ squash/carrots/ parsnips/celeriac/ swede/beetroot)	Grains (quinoa/pearl barley/ mixed grains)	Fresh herbs (mint/coriander/dill/ parsley/basil/sage)
Delicate greens (spinach/ watercress/rocket/ peas/beans)	Pasta or couscous	Dairy (cream/yogurt/ soured cream/crème fraîche/butter)
Summer (peppers/tomatoes/ aubergines/ courgettes/ cucumber)	Drained and rinsed tins (chickpeas/beans)	Cheese (feta/Parmesan/ Cheddar)
Hardy veg (chard/broccoli/ kale/cauliflower/ fennel/cabbage)	Rice or any noodles	Bread (croûtons/toasted pitta/breadcrumbs)
Any type of mushroom	Meat or fish (cooked chicken/ ham/responsibly sourced prawns or fish)	Seasonings (citrus/soy/chilli oil/ capers/fish sauce/ pesto)
		Any toasted seeds or nuts

PUT AN EGG ON IT

If there's eggs in the house MOB, you're well on your way to a fresh little meal. These are our easiest ever egg recipe ideas for one. Each one is a banger. Try them all.

EGG AND CHIPS:

Preheat the oven to 180°C fan (200°C/400°F/Gas Mark 6). Cut 2 medium potatoes, skin-on, into wedges, drizzle with oil and season. Sprinkle over dried oregano if you have any and roast for 40 minutes or until crispy and cooked through. Once the chips are done, fry off 2 eggs. Serve with condiments of your choice and baked beans if you fancy. Fresh, fresh, fresh.

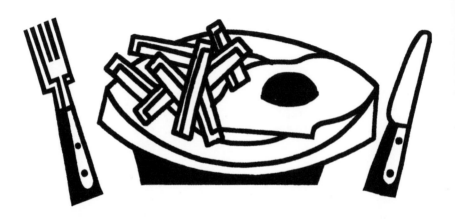

BUDGET SHAKSHUKA:

Fry off 1 sliced red or white onion in olive oil until soft. Add 1 chopped garlic clove, chilli flakes or chopped fresh chilli and cook for 1 minute more. Tip in a can of chopped tomatoes or passata. Add seasoning, along with a splash of any vinegar. Bubble away for 5 minutes, then crack in 2 eggs. Put on a lid and cook for 6–8 minutes until the whites are set but the yolks are still runny.

Optional extras: Dollop of yogurt, any fresh herbs chopped and scattered on top and/or a sliced pepper fried in with the onion. So good.

BOILED EGGS WITH CHEESE AND MARMITE SOLDIERS:

Preheat the grill to maximum. Boil 2 eggs for 4 minutes. Meanwhile, toast slices of any bread under the grill, then butter and spread one side of each with Marmite. Grate over whatever cheese you have. Slide under the grill until bubbling and browned, then cut into soldiers. Serve with the soft-boiled eggs.

RAPID RAMEN:

Boil 2 eggs for 6½ minutes until jammy – cool and peel. Make up your favourite packet ramen, dropping any green veg you have into the water with the noodles as they cook. Pile into a bowl. Top with the boiled eggs and drizzle over some hot sauce.

SPICED SCRAMBLE:

Fry off 1 chopped red or white onion until soft. Add 1 chopped garlic clove and some chopped ginger if you have any. Cook for 1 minute, then add spices of your choice (curry powder, chilli powder, cumin seeds, etc.). Fry for a further minute. Crack 2–3 eggs into a bowl, whisk and season. Pour the eggs into the pan and scramble with the other ingredients. Serve with rice.

USE-IT-UP FRITTATA:

There's not even a recipe for this one. Fry off any fillings you have in the house, add beaten eggs and seasoning – cook under a hot grill until completely set. 9 times out of 10 it's gonna taste like it came out of a restaurant kitchen. Try it.

IN YOUR CUPBOARDS

It's likely that you already have one or more of these ingredients in your cupboards. MOB, they're always there and sometimes they can seem a bit boring BUT... here are a few epic recipe ideas for you to freshen up the old staples.

BAKED BEANS

Fry off 1 chopped red or white onion until soft. Add 1 chopped garlic clove and some chopped ginger if you have any. Fry for 1 minute. Add whatever curry spices you have, then tip in a 400g (14oz) can of baked beans. Cook, stirring, until the beans are hot.

Optional extras: Make this go further by adding a can of coconut milk along with the beans to turn it into a curry. Top with fresh coriander if you have any. Also delicious stuffed into a wrap with fried paneer or yogurt.

TINNED TOMATOES

Fry off 1 sliced red or white onion. Add any spices you fancy (cumin, smoked paprika or ground coriander work well). Cook for 1 minute. Tip in a 400g (14oz) can of chopped tomatoes and a 400g (14oz) tin of drained chickpeas (keep the liquid, see page 95). Bubble away for 15 minutes, then stir in a leafy green veg such as spinach or kale. Once wilted, season to taste.

Optional extras: Fry chorizo and peppers off with the onion. Dollop over some yogurt and fresh herbs to serve. Eat with a piece of responsibly sourced fish.

PASTA

Fry off a few anchovies, loads of chopped garlic and chilli in a glug of olive oil until the anchovies have 'melted'. Toss in cooked pasta along with a splash of pasta water. Squeeze in some lemon juice and season with black pepper. Serve with shedloads of grated Parmesan. Ridiculously good.

RICE

Fry off 1 sliced red or white onion with a sliced pepper until soft. Add chopped garlic, smoked paprika and a good pinch of ground turmeric. Cook for 1 minute, then stir in 1 tbsp tomato purée and ½ mug of rice. Add 2 mugs of water or stock. Cook until the water has nearly all evaporated, then chuck in some frozen peas. Once hot, squeeze over some lemon juice if you have any.

Optional extras: Fry chorizo with the onion and pepper. Add cooked chicken along with the peas. Throw in whatever other veg you have in the kitchen.

SPICE IT UP

Getting to know your spices is the easiest way to make simple ingredients fly. Rather than always following a recipe. We're going to show you which fresh ingredients work well with each spice. This way, you won't have to follow a recipe every single time.

Ground cumin has an earthy, nutty and warm flavour that is great for roasting vegetables and when used as a seasoning for meat and fish.

Great with: *Yogurt, beetroot, butternut squash, lamb, spinach, carrots, halloumi, salmon, mackerel, chicken, sweet potatoes, cauliflower, fresh ginger, garlic, celeriac, mint, parsley, coriander, pomegranate, radishes.*

Smoked paprika is punchy, spicy and a little sweet, great for when you want to add some smoky depth to tomato-based meals and creamy dishes.

Great with: *Chorizo, onion, peppers, tomatoes, cod, prawns, sweet potatoes, chicken, pork, camembert, crème fraîche, soured cream, aubergine, courgette, mushrooms, parsley, garlic, lemon.*

Turmeric adds vivid colour and a slightly warming flavour. It is PERFECT for curried dishes and when mixed with a fat it provides a good coating for cheese and tofu before frying.

Great with: *Paneer, tofu, halloumi, chicken, spinach, broccoli, cauliflower, peas, lime, eggs, coriander, yogurt, fresh ginger, garlic, cabbage, Brussels sprouts, kale, chilli, parsnips, carrots.*

Dried oregano is a strong herb with an aromatic, warm and slightly bitter flavour that works well with summer ingredients.

Great with: *Sea bass, courgettes, tomatoes, feta, aubergine, lamb, potatoes, garlic, mint, basil, peppers, onion, olives, fennel, pork, lemon.*

Chilli flakes are very versatile, adding pops of heat and can be used for anything from spicing up a ramen to adding a kick to pesto.

Great with: *Chicken, cod, prawns, tomatoes, broccoli, yogurt, eggs, basil, garlic, fennel, celeriac, lamb, pork, butternut squash, kale, spinach, avocado, sweetcorn, green beans, chard, feta, Parmesan, asparagus.*

Ground coriander is aromatic and sweet with a slight citrus flavour. It cuts through fatty meats like lamb and pork and is great with stir-fried greens.

Great with: *Leeks, pork, lamb, kale, spinach, peas, chard, potatoes, onions, peppers, tomatoes, aubergine, chorizo, feta, sweet potatoes, butternut squash, garlic, fresh ginger, chilli, carrots.*

TAKE ONE ROAST CHICKEN

We need to get back to how they did it in the old days, viewing meat as more of a treat, eating less of it and then making the absolute most out of the leftovers. We're here to show you how one large roast chicken can go a LONG way.

PIE

Fry off sliced leeks and mushrooms until soft, then add mustard, garlic and crème fraîche or soured cream. Mix well, then fold through half the **cooked chicken**. Season to taste. Transfer to a baking dish, top with pastry of your choice and brush with a beaten egg. Make a steam hole, then bake at 200°C fan (220°C/425°F/Gas Mark 7) for 20 minutes until the pastry is crisp.

RISOTTO

Use the leftover **chicken stock** to make the most banging risotto – any flavour you want. Head to **www.mobkitchen.com** for all our risotto recipes.

ROAST CHICKEN

Put any visible leftover meat into an airtight container in the fridge. Keep the carcass for making stock (see right).

STOCK

If you stuffed the chicken with any herbs, lemon or garlic then keep them inside and put the **chicken carcass** into a large saucepan. Cover with water. Bring to the boil, then leave to simmer over a gentle heat for 2 hours, topping up with more water if you need to. Cool slightly, then pour the stock into a container.

→

MINESTRONE

Use half of your homemade chicken stock to make a minestrone soup. Simply fry off onion + carrot + celery until soft, add **chicken stock** and bring to a simmer, then tip in a handful of pasta. Once cooked, stir through half the **leftover cooked chicken** along with any leafy greens. Season to taste, adding a squeeze of lemon juice. Serve with grated Parmesan. Keep the remaining stock in the fridge.

TAKE TWO ROAST BUTTERNUT SQUASH

One for the veggie MOB. A couple of skin-on, roast butternut squash cut into chunks makes a brilliant base which can be turned into many meals for one.

TACOS

Fry off some chopped garlic with a little chilli powder, smoked paprika and ground cumin. Add a 400g (14oz) can of mixed beans and a few cherry tomatoes. Cook for 5 minutes. Season to taste. Make salsa and/or guac. Reheat ¼ **roast squash chunks**, then build your tacos inside tortilla wraps.

ASIAN SALAD

Cook a portion of your favourite rice or grains. Mix the juice of ½ lime with a thumb-sized piece of fresh grated ginger, 1 tsp soy sauce and 1 tbsp peanut butter. Stir together, adding enough water to make a loose dressing. Toss the rice/grains in the dressing, then fold through ¼ **roast squash chunks**, any mixed herbs you might have and some crunchy raw veg – cucumber, radishes or pepper work well.

ROAST BUTTERNUT SQUASH

Put the roast squash chunks in an airtight container in the fridge.

PASTA SAUCE

Blitz ¼ **roast squash chunks** with some warm water, a splash of lemon or vinegar and a small garlic clove until smooth. Cook the pasta for 2 minutes less than the packet instructions say, then drain. Mix the pasta with butternut sauce over the heat. Top with feta or capers and some peppery rocket.

CURRIED SOUP

Fry off your chosen curry paste in a pan, add ¼ **roast squash chunks** along with a 400g (14oz) tin of coconut milk. Bubble away for 10 minutes, then blitz. Season to taste.

MEAL PREP BOSS

Making one big batch of food for the rest of the week helps manage your food waste. But the same meal tends to get boring by day three. To resist the temptation of nipping to your local sandwich joint, we're going to show you how to keep those one person meals fresh all week.

BOLOGNESE

A MOB classic. If we are going to eat a little high-quality beef, then making it last over the course of a week is the best way to really enjoy this treat and make it as sustainable as possible.

Bolognese with sweet potato mash
Chop 1 skin-on sweet potato into chunks. Put it into a microwaveable container with ½ lime. Cook for 8–10 minutes until completely soft. Carefully remove the lime, squeeze all the juice over the potato, then mash. Season to taste. Can be made the night before and then reheated and served with one portion of bolognese.

Tacos
Make a salsa/guac of your choice. Serve with the remaining chilli con carne and some hard taco shells or tortillas.

The ultimate chilli con carne
Heat some oil in a saucepan. Add spices of your choice (we like chilli powder, ground cumin and smoked paprika). Cook for 30 seconds, then tip the remaining two portions of bolognese into a saucepan. Add a 400g (14oz) can of drained beans of your choice and a square of dark chocolate if you have any. Cook for 5–10 minutes, seasoning and adjusting the spice to your taste. Can be made the night before and reheated – save one portion for the following day.

CURRY

Be it veggie or otherwise, we know that the MOB LOVE a cuzza. These are our hacks for keeping it fresh.

Ramen
Heat a portion of curry in a saucepan. Add enough water to make the curry into a soup consistency, then drop in your noodles. Simmer until the noodles are just cooked. Serve with hot sauce or chilli oil if you like. Can be made the night before and then reheated.

Curry chickpeas
Mix the remaining two portions of curry with a 400g (14oz) tin of drained chickpeas (keep the water, see page 95). Save half for tomorrow, then mix the rest of the curried chickpeas with a dollop of yogurt and some mango chutney/fresh coriander if you have any. Reheat to eat.

Paneer or halloumi wrap
Mix a splash of oil of your choice with some ground turmeric. Chop some paneer or halloumi into cubes. Toss the cheese with the turmeric and fry until bright golden and crisp. Divide the remaining curried chickpeas between wraps along with the fried cheese and some spinach leaves. Boom.

ROAST VEGGIES

One of the simplest, cheapest and most sustainable forms of meal prep. A pile of roast veggies can set you up.

Pesto pasta
Cook off a portion of any pasta. Blitz any fresh herbs you have with 1 small garlic clove, olive oil and any nuts or seeds to make a pesto. Mix pasta and ½ pesto with a portion of roast veggies. Tastes good hot or cold the next day.

Tart
Cook off a sheet of puff pastry at 180°C fan (200°C/400°F/Gas Mark 6) for 20 minutes until crisp. Spread over the remaining pesto, top with a portion of roast veg and anything else you may fancy. Make the night before for lunch for the next two days.

Cheat's parmigiana
Preheat the grill to medium. Pour some passata into a roasting dish. Add the remaining roast veggies and some chilli flakes. Season and mix together, then tear over some mozzarella and any bread you may have. Grill for 5–10 minutes until the cheese is melted and bubbling and the bread is crisp.

FOOD WASTE

Now we've covered cooking, it's time to look at food waste.

This chapter is all about providing you with the techniques and information you need to really maximize the value of your food. This is where we need to hone our kitchen skills even further by making use of parts of ingredients that we may have otherwise thrown in the bin.

Did you know that 1/3 of all food produced globally is wasted[61] and that 70% of food waste comes from households?[62] To tackle this, we've been asking around to find out what ingredients are the most wasted and what we can do to help you save them.

Then there's the question of how to store your food in the first place. Have you ever wondered how long certain foods are 'good' for? Well, we've got the answers...

DATE DETECTIVE

Before we begin tackling food waste, let's delve into the ever-confusing sell-by, best before and use-by dates. Have you ever felt torn as to whether you should throw something away, as it was definitely 'out of date' but looked fine?

We have, so we wanted to find out exactly what these dates actually mean. When is it important to stick to them and in what situations is your common sense the best approach?

Sell by: This is a date intended for retailers and can be largely ignored by us. It is a guideline for shops to know that the food being sold will be the best quality before this date. It is often seen on ingredients such as baked beans or crisps. Eating the food after this date is fine. Typically, an item has 1/3 of its shelf life left after its sell-by date.[63]

Best before: This date is all about quality. Eating the food before this date means that it will be at its tastiest. It is not unsafe to eat something past its best before date.

Use by: This is the most important date because it concerns food safety. It is the official guideline for when the food should be eaten. At MOB Kitchen, we always stick to the use-by dates on any raw meat or fish. However, it is worth mentioning that these dates often err on the side of caution, so if in doubt, use some common sense.

COMMON SENSE

This is our questionnaire to help you decide whether to eat food past its use-by date, which will hopefully prevent you from chucking away something that's still good to eat. Get checking:

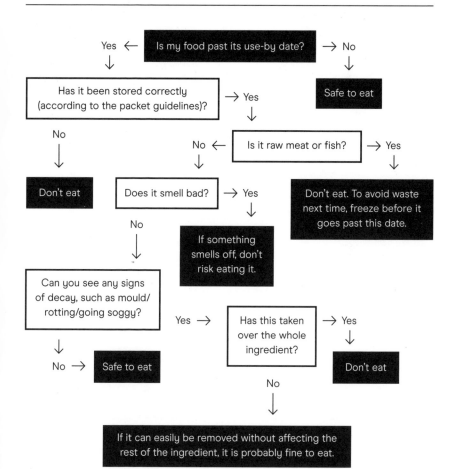

Yes ← **Is my food past its use-by date?** → No

↓ (Yes) → Has it been stored correctly (according to the packet guidelines)?

↓ (No, from use-by) → **Safe to eat**

Has it been stored correctly (according to the packet guidelines)? → Yes ↓

No ↓ → **Don't eat**

Is it raw meat or fish? ← No ↓ / → Yes ↓

Does it smell bad? → Yes ↓

→ Yes: **Don't eat. To avoid waste next time, freeze before it goes past this date.**

If something smells off, don't risk eating it.

No ↓

Can you see any signs of decay, such as mould/rotting/going soggy?

Yes → Has this taken over the whole ingredient? → Yes ↓

Don't eat

No → **Safe to eat**

No ↓

If it can easily be removed without affecting the rest of the ingredient, it is probably fine to eat.

FOOD STORAGE

Our top storage hacks will help you keep your food fresher for longer.

IN THE FRIDGE

Always check that the fridge is at the right temperature. If it's too cold then things may be freezing at the back of it – this is a killer for green leaves or herbs.

Cheese: Remove from any plastic packaging as this will make it sweaty. Store in an airtight container. Key tip.

Raw meat and fish: Keep inside sealed packaging, then once open transfer to an airtight container to keep fresh, making a note of the use-by date.

Soft herbs: Keep in their packaging or wrapped in a clean tea towel. When looking a little limp, keep upright with the stems in a jar of cold water.

Berries: Remove from their packaging as they can often get squashed. Keep in a bowl. Important note – remember to eat any berry that is looking overripe as one can make the rest of them turn.

Green leaves: Keep in their packaging. Once open, insert a clean dry cloth into the bag. It will absorb excess water and stop them from going off as quickly.

Mushrooms: Keep under a tea towel or in a dark paper bag.

Cut veg: Keep the remaining part in a jar filled with a little cold water, cut-side down.

Woody herbs: Keep in their packaging in the fridge.

OUT OF THE FRIDGE:

Some ingredients are actually better stored at room temperature than in the fridge.

Tomatoes: Remove from any packaging and keep in a bowl. If they are slightly hard, put the bowl on a windowsill to ripen. Never store tomatoes in the fridge as they lose their flavour.

Citrus fruit: Take out of any packaging as this will make them go off quicker. Instead keep in a bowl.

Bread: Keep in a dark, cool place, such as a cloth bag or opaque airtight container.

Garlic and onions: Keep out of the fridge in a cool, dark place away from other ingredients as their flavours can spread.

Bananas: In the fridge they blacken, so keep at room temperature away from other fruit as they will ripen that quicker.

Potatoes: Store your potatoes in a cool, dark place out of their original packaging in a cloth bag or cupboard to stay fresh.

Butternut squash/swede/celeriac: All large robust root veg should be stored in a cool place out of direct sunlight.

IN OR OUT?:

Eggs: Can be kept in or out of the fridge – it's important to choose one or the other and then stick to it as eggs will stay fresher when kept at a constant temperature.

Key cooking tip – room temperature eggs are easier to cook with, especially when working with recipe timings such as for soft-boiled eggs.

Fridge eggs will keep for slightly longer, but make sure they are kept away from strong smells as their shells are porous and will absorb them.

Avocados: Unripe avocados should be kept out at room temperature, whereas super ripe ones are best in the fridge. If you have a really hard avocado that you want to eat, whack it next to some bananas which will soften it.
If you only eat half an avo, put the rest in the fridge with the stone in. Keeping the stone in will help stop the avocado from turning brown.

THE FREEZER IS YOUR FRIEND

When it comes to stopping food waste, the freezer is your saviour, be it saving a portion of something you have already cooked or rescuing ingredients from going off. Plus, ingredients such as frozen peas and berries make great cooking staples, and as they are picked at the height of their season you can enjoy them at their best all year long.

What to freeze:

Nearly all ingredients can be frozen. The only things you shouldn't freeze are fruit and veg with a high water content such as cucumber, apple, celery, cabbage or salad leaves as these tend to turn mushy when defrosted.

How to freeze like a pro:

1. Reusable freezer bags. Decant bulkier items into freezer bags. Even mince can be squashed into a freezer bag and then frozen in a flat lump to make room for more ingredients. Bags are especially good for freezing sauces.

2. Portion. If you've made four portions of a pasta bake and you are only going to eat two, then freeze the other two in separate portions so you can get one at a time out to eat.

3. Label. Write what the food is you've frozen and the date you froze it. You should try to eat up food in your freezer every 3 months, so having the date shows you what to use up first.

4. Defrost properly. Get things out of the freezer the day before and leave in the fridge over a bowl to catch any water as it defrosts.

Things we bet you didn't know you could freeze:

→ Hummus – defrost in the fridge before eating

→ Egg whites (see page 95)

→ Herbs (see page 91)

→ Chilli – grate from frozen

→ Ginger – grate from frozen

→ Pesto and sauces – use from frozen

→ Bananas (see page 90)

→ Bread (see page 92)

→ Milk – defrost in fridge and shake well before using

→ Butter – use from frozen

→ Grated cheese – use from frozen

→ Potatoes (see page 93)

→ Cooked rice and pasta (see page 101)

CONDIMENT CITY

Ever looked into your fridge and been overwhelmed by the amount of half-used condiments? We've got some ideas for you on how to use them up.

Chipotle paste

→ Whisk into eggs for an omelette.
→ Stir a spoonful into a can of baked beans for a spicy hit.
→ Use to marinate chicken or white fish.

Dijon mustard

→ Stir into cheese sauce.
→ Add a spoonful to your mash.
→ Use the very dregs of the jar to make a dressing. Pour in olive oil, lemon juice, a pinch of sugar and some seasoning, then shake the jar to combine. So simple.

Cranberry sauce

→ Spread on toast for cheese toasties with a jammy twist.
→ Mix with maple syrup for a sweet topper for pancakes, waffles or granola and yogurt.
→ Glaze roast pork with cranberry sauce for the final 10 minutes of cooking.

Pesto

→ Freshen up cooked greens with a spoonful of pesto stirred through.
→ Mix with soft dairy (like crème fraîche, yogurt or cream cheese) for a creamy gnocchi sauce.
→ Coat torn, stale bread in pesto and olive oil and bake at 160°C fan (180°C/350°F/Gas Mark 4) for 10 minutes to make croûtons – this is our favourite.

Curry paste

→ Combine with oil, then use to coat cauliflower before roasting.
→ Mix with a tin of coconut milk and lentils for a quick dahl.
→ Mix with yogurt and use to marinade halloumi, meat or salmon.

GO BANANAS

This section is crucial, because bananas are one of our most wasted foods.[64] Now we know how to store them (out of the fridge away from other fruit), here are our top hacks to stop us from wasting them.

→ **Freeze.** Peel the banana, roughly slice and then freeze it. Perfect for smoothies, porridge or banana ice cream.

→ **Banana 'ice cream'.** It couldn't be simpler. Blitz up your frozen banana pieces. Be patient, first it will turn from crumbly to gooey then into a creamy soft-serve ice cream texture. And that's it, vegan ice cream. Couldn't be simpler if it tried. Experiment with adding different flavours before you blitz – a spoonful of peanut butter or cocoa powder are fresh.

→ **Banana pancakes.** Mash 1 large, ripe banana into a mixing bowl. Crack in 1 egg. Add 1 mug self-raising flour and 1 mug milk of your choice. Whisk everything together to form a batter. Get a non-stick pan over a high heat. Drizzle in a little oil, then spoon in pancake-sized amounts of batter. Fry the pancakes for 1–2 minutes on each side. Easyyy.

GO GREEN

Herbs are constantly at the top of our most wasted list. Here are our hacks to stop that from happening:

WOODY HERBS:

Rosemary, thyme and bay leaves all freeze perfectly. You can then chuck them into whatever you are making straight from frozen to add extra flavour.

FRESH HERBS:

Time to turn to the sauce. Turn whatever mixture of leftover fresh herbs you have into the freshest, flavour-packed sauces and salsas. Once made, the green sauces will last for 2–3 days in the fridge and can be used to elevate pretty much any dish. Salads. Soups. Stews. Anything.

Salsa verde
Finely chopped mixed herbs + chopped capers + spoonful Dijon mustard + squeeze of lemon + glug of olive oil + seasoning = done.

Chimichurri
Blitz mixed herbs + 1 small garlic clove + good pinch dried oregano + ½ small onion + glug of olive oil + ½ chilli + good splash of red wine vinegar + seasoning.

Aji verde
Blitz mixed herbs + dollop of mayo + a few pickled jalapeños from the jar + 1 small garlic clove + squeeze of lime + seasoning.

Zhug
Blitz mixed herbs + big pinch chilli flakes + ground cumin + ground coriander + glug of olive oil + splash of white/red wine vinegar + seasoning.

Green chutney
Blitz mixed herbs + 1 small garlic clove + small piece of ginger + dollop of yogurt + squeeze of lemon + garam masala + seasoning.

BREAD

This stat is mind-blowing – we waste a staggering 24 million slices of bread every day[65] and in the UK alone, 44% of the bread produced is never eaten.[66]

The good news is that stopping wasting bread couldn't be simpler:

1. **Freeze.** Slice bread if it isn't already, seal inside a reusable freezer bag and pop in the freezer. Toast bread straight from frozen – most toasters now have a defrost button and if they don't just toast low and slow.

2. **The crunchiest croûtons.** If your bread is already stale, cut into chunks, drizzle with olive oil and roast at 180°C fan (200°C/400°F/Gas Mark 6) until crisp. These are great in a salad, or blitz the croûtons into fine breadcrumbs and keep in an airtight container for up to two months to use for making homemade fish fingers or chicken nuggets.

3. **Pud.** Butter a roasting dish, then add 8 slices of semi-stale bread along with some dried fruit if you have any. Crack 3 eggs into a bowl. Pour in 1 mug of milk and ½ mug double cream, then add 3 tbsp sugar. Whisk together into a smooth custard. Pour the custard over the bread slices, then sprinkle over a little more sugar. Bake for 30 minutes at 160°C fan (180°C/350°F/Gas Mark 4) until puffed and golden. Delicious.

POTATOES

We may love spuds but according to waste polls we are not eating them fast enough. The UK throws an estimated 5.8 million potatoes into the bin every day.[67]

This is how we tackle it:

1. **Share.** Most potatoes come in a big bag, so share them with your housemates and you will be less likely to waste any.

2. **Store potatoes correctly.** In the dark, out of the way of other vegetables.

DID YOU KNOW YOU CAN EVEN FREEZE COOKED CHIPS AND THEN REHEAT IN A SUPER-HOT OVEN? SIMPLE.

3. **Cut off the bad bits.** For slightly green and/or sprouted potatoes, just slice off the bad bit and use the rest.

4. **Use the peel.** If a recipe calls for the potato to be peeled, keep the peelings in your fridge and use to make homemade crisps (see page 98).

5. **Pre-cook.** Cooking potatoes can take a while, so our advice to you MOB is to pre-boil a bunch of them for 10–15 minutes until just tender, then keep in your fridge for 3–4 days. When ready to eat, cut the pre-cooked potatoes into wedges, drizzle with oil and add seasoning. Roast at 180°C fan (200°C/400°F/Gas Mark 6) until crisp.

6. **Mash it up.** If the potatoes are about to turn, boil them all and mash. Scrape mash into a reusable freezer bag, label, date and freeze for another time. Defrost in the fridge before using.

DID YOU KNOW YOU CAN KEEP…?

PARMESAN RINDS

Throwing away Parmesan rinds is the sin of all sins. Once you've finished eating the cheese freeze the rind. It is a gift from the culinary gods.

Grated Parmesan makes all food taste delicious. Pasta ✓ Risotto ✓ Soup ✓

And all of that flavour is also in the rind. So, simply chuck the frozen rind into a pan when you are making a tomato sauce, a meat ragù, stews, soups or even heating up baked beans. The rind will begin to melt and impart its Parmesan-y flavour to whatever you are making.

It isn't just Parmesan. Any hard cheese rind can be frozen and used in the same way.

FETA

Our top hack for keeping feta even longer is to marinate it. Just whack it in a jar and cover with a bit of olive oil and any other flavours you want (fresh herbs/garlic/chilli, etc.). Not only does it last longer this way, it tastes sublime. Once the feta is finished keep the oil and start again.

EGGS

Have you ever separated an egg because you've needed the yolk for the ultimate carbonara and not known what to do with the white? Or, you've needed the whites for a meringue but not the yolks? Well, the good news is you can keep both of them and use them both separately. Here's how:

Egg whites:
Freeze really well in a reusable bag. Label and date, then keep in the freezer for up to 3 months. Defrost overnight and use in a meringue recipe or to froth up cocktails, such as a rum sour.

Egg yolks:
Keep covered in the fridge for 1–2 days. Mix with a little milk and use to brush pastry. Use to make custard, a carbonara or whisk into cracked whole eggs for a rich and creamy scramble.

The egg test:
Eggs have their own unique way of letting us know if they are safe to eat, even when they may be technically out of date. To test if an egg is fresh, place it in a bowl of water. If the egg sinks – eat. If it floats – don't.

CHICKPEA WATER

As weird as it sounds, you're gonna want to save your chickpea water.

Why?

Well, someone, somewhere did some experimenting and found that the aquafaba (chickpea water) acts pretty much like a vegan egg replacement and can be used to make vegan mayonnaise and even meringues.

Look online for the best recipe. Our top tip when making either is to have patience. Whisking up aquafaba for meringues takes twice as long as regular egg whites – persist and it will happen.

Drain the liquid from a can of chickpeas and keep in the fridge for up to a week. Adding a bit of this liquid to a food processor when making hummus also makes it super creamy. Who knew?

ALL OF IT

One of the easiest ways to stop food waste is by eating the whole vegetable. That may sound blindingly obvious, but when you stop and think about it, we rarely do it.

STEMS AND STALKS

Let's start from the bottom. There is no need to be throwing away vegetable cores or stems. All you need to remember is that this part is tougher, so it requires different preparation from the rest of the vegetable.

Cauliflower and broccoli stems:
These are absolutely delicious, with a really nutty flavour. Thinly slice and roast with the florets. If cooking on the hob, chuck the sliced stems into the pan first and cook for 2–3 minutes before adding the florets.

Robust greens:
Kale, cavolo nero and chard all have thick stalks which taste good but just need longer cooking. Slice down either side of the stalk to remove the leaves, then finely slice the stalk. Fry these off first until soft, then add the leaves.

LEAVES AND TOPS

Cauliflower leaves:
It's controversial but these are the best bit. Chuck into soups and stews or add to roasted florets for the last 10 minutes of cooking. We love them roasted and coated in a mix of curry spices. Boom.

Beetroot leaves:
These are more delicate than the root and taste wonderful stir-fried. Make sure you give them a good wash, then fry off with any spices you fancy.

Carrot tops:
If you ever buy carrots with their tops on, you can make the best pesto:
Drop the carrot tops into a pan of boiling water. Cook for 1 minute until wilted and vivid green, then drain and cool. Blitz in a food processor with any toasted nut you fancy + glug of olive oil + 1 small garlic clove + squeeze of lemon + grated Parmesan if you have it + seasoning.

Spring onion and leek tops:
There is no reason for us not to be eating the green part of these vegetables, just make sure you wash them well before preparing as dirt tends to cling to the leaves. Use in exactly the same way as the rest of the veg, in soups, risottos and other dishes.

HERBS

You can eat the stalks of pretty much all soft herbs, excluding mint (which can be used in tea, see right). Simply blitz along with the leaves or roughly chop when adding the herbs to dishes.

Herbal teas:
Making your own herbal teas couldn't be easier. Boil the kettle, put your leftover herbs and any other ingredients into a jug, then pour over the water and leave to brew. It is a game changer.

These are our favourite flavours:

→ **Rosemary stalk and lemon skin:**
   2 rosemary stalks + 1 used lemon + boiling water. Brew for 2–3 minutes.

→ **Strawberry tops and lime skin:**
   Green tops from 1 punnet of strawberries + ½ used lime + boiling water. Brew for 5 minutes.

→ **Lemongrass and ginger:**
   1 outer lemongrass layer + thumb of ginger peelings + spoonful of honey/maple syrup + boiling water. Brew for 5 minutes.

→ **Mint stalks:**
   Handful of mint stalks + pinch of sugar (if you like) + boiling water. Brew for 2–3 minutes.

VEGETABLE SKINS

One basic rule MOB: if you don't need to peel it, then don't. Not only are most nutrients in the skin of fruit and vegetables, but saving these peelings to make crisps or veg stock cuts down on food waste dramatically.

Veg crisps:
No need to spend your money on a bag of crisps. Make your own.

Preheat the oven to 180°C fan (200°C/400°F/Gas Mark 6). Toss your veg peelings in a good drizzle of olive oil + any spices you fancy + salt and pepper.

Spread out into a single layer on a large baking tray. Roast in the centre of your oven for 8–10 minutes. Start checking the crisps after 5 minutes, as some of the ones around the sides of the tray will cook quicker than others.

Keep an eye on them, removing from the tray when crisp. These will keep for up to 3 days in an airtight container.

Flavour suggestions:
→ curry powder
→ smoked paprika and ground cumin
→ garlic powder and dried oregano
→ sea salt and malt vinegar

Potato peelings will take slightly longer to crisp in the oven – give them 20–25 minutes or until crisp. Start checking them after 15 minutes, removing any that are cooked.

Veg stock:
We use stock in so many recipes at MOB Kitchen, yet before we started our journey to be more sustainable we never made our own. It's unbelievably easy to make. It reduces food waste, it's much more flavourful AND it's package free!

This isn't something you need to be doing every day. Keep a reusable container in the fridge or freezer to store your veg scraps. Once full, it's time to get on the stock prep.

The key is to not have too high a water-to-veg-scraps ratio, otherwise the stock will end up being diluted and weak. All you need to do is tip your veg scraps into a saucepan and pour in cold water, so it just covers the vegetables.

Bring to the boil, then simmer for an hour. Sieve the stock, smooshing all the last bits from the veg with a ladle. Keep the stock in the fridge or reduce further until concentrated and freeze in an ice cube tray to add to sauces, soups and stews straight from frozen.

What veg can I use?

Yes:
→ Onion and garlic skins
→ Stalks and stems from herbs
→ Celery ends
→ Any forgotten shrivelled mushrooms
→ Carrot and parsnip tops

No:
→ Potato peelings are too starchy and will make the stock cloudy – use for crisps instead (see page 97)
→ Citrus fruit as the flavour will be overpowering – stick to using in tea (see page 97)
→ Cruciferous vegetables such as cabbage or cauliflower

GROW

YOUR OWN

MOB, this tip is key. You can regrow your own vegetables from scraps you have in your kitchen – nothing is more satisfying. Fennel, celery, spring onion, lettuce and leek work best.

It does require some love and care. We can't work miracles, so you're going to have to tend to those sprouting beauties to keep them alive – but in theory it's pretty magical. All you need to do is follow these easy steps:

1. Cut off the base of the vegetable, keeping on any roots that may be attached.

2. Put the base into a glass jar or shallow container, then pour in enough water to just cover.

3. Leave on a sunny windowsill.

4. Top up the water when needed.

5. Watch the veg scraps grow. Within a week you should start to see shoots forming!

6. Transfer the new plant to a pot of soil and keep loving and caring for it until you've got a whole new vegetable.

YOU'VE MADE TOO MUCH...

It's hard to always get your carb portions right, especially if you are cooking for one. The easiest way to avoid this is weighing out the amount of pasta or rice you need, but if you have made too much then don't fret as we've got some solutions for you.

RICE

If you accidentally cook more rice than you need (it happens 9 times out of 10), as soon as it has cooled, put it into an airtight container and store in the fridge. Cooked rice will last for up to 4 days in the fridge.

You can also freeze cooked rice. To keep it fluffy when you reheat, put the hot cooked rice into a container/reusable freezer bag and seal – this will trap any steam. Once cool, label, date and then keep in the freezer for up to 1 month.

Reheat straight from frozen in the microwave, making sure the rice is piping hot before eating.

Recipe ideas for leftover rice:

→ Egg fried rice works best with cold leftover rice. Cold rice + a super-hot pan = extra crispy bits. Easy.

→ Speedy pilau. Fry chopped onions with garlic, ginger, chilli and spices until soft. Add leftover rice with a big splash of water and stir to reheat. Serve topped with yogurt and with a piece of meat or fish if you like.

PASTA

Cooked pasta will keep in the fridge for up to 3–5 days. To keep it at its best, toss plain cooked pasta in a little olive oil to stop the pieces from sticking together, then place in an airtight container. Pasta already coated in sauce will last in the fridge for up to 2 days.

If freezing, mix the cooked plain pasta with a little oil in a reusable freezer bag – label, date and use within 1 month. Make sure to seal the bag correctly before putting it into the freezer, otherwise the pasta can go mushy when reheated.

Reheat straight from frozen by dropping into a pan of boiling water and cooking for 30 seconds until the pasta is piping hot.

Recipe ideas for leftover pasta:

→ Pasta salad. We like ours with a mustardy dressing and peas.

→ Spaghetti cake – this one is a winner. Crack 3 eggs into a bowl, season with salt, pepper and grated Parmesan. Stir through any cooked long pasta. Melt some butter in a frying pan, add 1 chopped garlic clove and cook for 1 minute. Tip in the spaghetti and egg and flatten out in the pan. Fry for about 5 minutes on each side until golden brown. Cut into wedges to serve.

Now you're clued up on how to shop, cook consciously and tackle food waste, an area we haven't really touched on yet is arguably the world's most important resource – water. In this chapter we look into why water is so valuable for our food and how we can be better at saving it.

WATER

Put simply, water = life.

Water is needed to keep any living thing, be it plants, animals or us, alive. Human beings can survive for around 3 weeks without food but only 3–4 days without water.[68]

Clean water is needed not only for drinking and sanitation but also to produce our food.

With water so readily available to us in the West, it's important to remember that everyday things we take for granted like flushing a toilet, having a wash or even turning on the tap for a drink are, in fact, a luxury.

Here are the facts:

→ **29%** of the world's population do not have access to safely managed drinking water.[69]

→ According to the World Health Organization (WHO) **263 million people** spend more than **30 minutes** per trip to collect water.[70]

→ Women are responsible for water collection in 8/10 households.[71]

→ **1.8 billion people** around the world drink untreated waste water. This is water that has already been used by humans, often from the toilet and can be contaminated.[72]

→ **60%** of the world's population don't have access to a safely managed toilet.[73]

→ A recent WHO report claims that by 2025, **1 in 4** of the world's children will be living in areas of extremely high water stress – locations in which there is not enough water to meet the human demand.[74]

FOOD FOR THOUGHT

This next fact is incredible:

To produce enough food for one person's recommended daily calorie intake requires 2000–2500 litres (440–550 gallons) of water. That's the equivalent of 16 baths per day.[75]

So, how much water does it take to grow a single ingredient? Here's a list of some of the world's thirstiest foods.[76]

For every 1kg (2lb 4oz) food	Litres of water needed
Almonds	16,098
Avocado	1,176
Beef	15,415
Butter	5,553
Cheese	3,178
Chicken	4,325
Chickpeas	10,148
Chocolate	17,196
Coffee	18,900
Lentils	5,875
Oats	2,420
Olives	3,025
Pork	5,988
Rice	2,497
Soy	2,145
Wheat (pasta)	1,830, 1,847
Wheat (bread)	1,608

SAVING WATER

We have identified all of the areas in the kitchen where water is wasted and come up with some great tips on how to cut back on that waste – check it out:

1. **Defrosting.** Keep your food in the fridge overnight or defrost in a microwave rather than using hot water.

2. **A cup of tea.** How many people are having a cuppa? Fill the kettle to the amount you need. If it's only you, it's a waste of water and electricity to boil a full kettle.

3. **Cold water from the fridge.** We all love drinking cold water, so instead of waiting for the tap to get cold, fill up a jug as soon as the water starts running and put it in the fridge to chill. This is key.

4. **Use the right size pan.** Sounds obviou, but using the right size pan for one person = less water.

5. **Cook a one-pot meal every week.** Using one pan once a week for your dinner is so easy. Less washing up = less water. Plus, if everything is being cooked in one pan you won't be using water to cook something elsewhere.

6. **Swap boiling.** Some things have to be boiled, but wherever you can, try and swap it for a cooking method that saves water such as frying, steaming, roasting or grilling.

7. **Double use.** Boil multiple ingredients in the same cooking water, dropping them into the pan according to how long they will take, such as noodles and eggs or rice and kale.

8. **Plug the sink or buy a basin and keep it old school.** Be it washing veg or doing the dishes, fill the sink with the amount of water you need, rather than washing under running water.

9. **Leave stubborn food to soak.** If it's not coming off, then leave it to soften rather than trying and failing to wash the pan under running water.

10. **Dishwasher.** Only put it on when it is COMPLETELY full. Scrape rather than rinse the plates. Use an eco-cycle which uses less water.

KITCHEN SINK

A guide on how to get the best out of your washing up:

→ Scrape and stack the plates to one side. The cleaner the plates are before you start, the longer you can use the same water for washing. Organization is key. It takes way longer to wash up if everything is dumped into the sink dirty.

→ Washing up liquid. When you are down to the dregs, fill the bottle with a little water. You'll be surprised how much longer it lasts.

→ Soap bar. Don't underestimate using a humble plastic-free bar of soap for washing your hands.

→ Make sure you have a full sink of water before you start cleaning.

→ When your cloths get dirty, don't bin them (we all do it). Put them in the washing machine along with some dark clothes and reuse.

→ Paper towels. Ditch them and use cloths or a tea towel instead.

PLASTIC

We are nearly at the end of our journey, there's just one thing left – our ever-growing, gigantic pile of rubbish.

The whole out-of-sight-out-of-mind approach doesn't really work now that we've got the internet and Sir David Attenborough showing us how our plastic ends up being mistaken for food, and fed to baby albatrosses on the remote island of South Georgia...

It's a really hard one because it's virtually impossible to go totally plastic free. But what we can do is be more conscious of our plastic consumption, especially when it comes to how we manage our rubbish in the first place.

Before writing this book, we had no idea how difficult it is to recycle properly. In this chapter, we'll walk you through it, show you the differences between types of plastic and help you to minimize your own single-use plastic consumption with some simple swaps.

PROTECT OUR OCEANS

We are now at the point where pieces of plastic have been found almost everywhere on earth, from microplastics at the bottom of the Mariana Trench, to the mountain peaks of the French Pyrenees.[77] For each Instagrammed picture of an idyllic unspoiled beach, turn the camera to another angle and you'll find a tangle of washed-up plastic.

The facts are devastating:

→ It is estimated that there is already **150 million tonnes** of plastic in the ocean. That's the same amount as if the ocean contained 25 million killer whales.[78]

→ **1 rubbish truck** of plastic enters the ocean every minute.[79]

→ Over **90% of seabirds** have plastic +++in their stomachs.[80]

→ All **7 species** of turtle are known to get entangled in plastic.[81]

→ There are **5 massive trash islands** in the ocean: the one between California and Hawaii is the same size as Texas.[82]

→ If our level of consumption continues unchecked, by **2050** there could be more plastic in the sea than fish (by weight)[83].

THE FACTS

In short, our obsession with single-use plastic is polluting our planet. All of this damage has only happened in the last 50–60 years, with half of all plastics ever manufactured made within the last 15 years.[84]

So, what's the deal with it?
Plastic is a man-made material produced from naturally occurring materials like coal and oil. It is extremely versatile. You can pretty much make anything you want out of plastic. It's also cheaper to make a new plastic bottle than a recycled one.

The most depressing part is that we now know it takes 450 years for a single plastic bottle to break down.[85] But despite growing environmental pressure, oil-led plastic production is predicted to grow.

Microplastics – what are they?
Tiny pieces of plastic smaller than 5mm. Some are designed this way as abrasives for cosmetics (microbeads), while others are fragments that have fallen away from larger plastics.

Where can they be found?
When we think of microplastics, we most commonly think of the ocean where fish and marine animals unknowingly eat them along with their food. Research done by the WHO (World Health Organization) has found that microplastics are actually present in all bodies of water, including the water we drink.[86]

Microplastics in our water supply = microplastics in our food.

What does this mean?
We don't yet know the full effects that consuming this plastic has on our health, but scientists agree that it is definitely not natural. As microplastics are so small there's also no real way of telling how much we are ingesting.

So, while getting rid of plastic seems like an absolute no-brainer, unfortunately it isn't that simple.

We use plastic for so many helpful things, from hospital equipment to airbags in cars. If we were to get rid of it completely, we would need to replace one man-made material with another.

What can we do?
Say no to single-use plastic where you can. If you don't need it, don't buy it.

SINGLE USE SWAPS

These supermarket food swaps will help you ditch single-use plastic.

PLASTIC WRAPPED   SWAP FOR

Oil and vinegars	↔	Buy glass bottles instead
Bread	↔	Pick bread up from the bakery section of the supermarket in a tote or paper bag
Fruit and veg	↔	Buy loose
Spices	↔	Buy in jars (and use refill stores to replenish)
Condiments	↔	There's always an option to get your favourite condiment in glass
Tea bags (some brands contain plastic)	↔	Loose tea and tea bags without plastic (check online – a lot of supermarket own brand tea is good)
Soft drinks	↔	Buy in glass bottles instead

For more info on how to avoid packaging when shopping, see chapter 5 on Shopping and Organisation.

Reducing your single-use plastic consumption isn't going to happen overnight.

Try to think of it as a journey and use this as your check list:

→ What have I already changed?
→ What single-use plastic can I swap out now?
→ What can I plan to get rid of next?

SLING IT CLING

Clingfilm or plastic wrap has an interesting and unexpected history. It was first developed in 1933 as a type of spray-on plastic to be used on US fighter jets. It was later used as a lining for US army's boots.

In 1949, Saran Wrap was introduced to supermarkets in America, and by 1953 clingfilm was available to buy in the UK.

Now it is pretty much everywhere and for years it has been the go-to for keeping food wrapped up and fresh. But the huge problem is that clingfilm is NOT recyclable.

Cutting back on clingfilm is one of those situations where we're really going to have to change the habit. When you've been using something for years it's hard to think of an alternative.

So, how can we sack it off?

1. Put food into reusable containers instead.
2. Keep leftovers in a bowl and place a plate on top to cover.
3. Use glass jars for small amounts of leftovers.
4. Use a tote bag to store bread in.
5. Wrap sandwiches in paper or, better still, reuse paper that you've bought the bread in.
6. Beeswax food wraps are a reusable alternative to clingfilm and can be machine washed or hand washed in cold water to prevent the wax from melting. Avoid using beeswax wraps for raw meat and fish, as they can't be fully sanitized in hot water. Also avoid using with pineapple as it contains an enzyme that corrodes the protective surface of the wraps.

REUSE AND RECYCLE

Even if you take up all of the approaches listed on the previous pages, it's inevitable that you are going to generate some rubbish. Rather than beat yourself up about it, here's how to repurpose your waste.

HERE'S OUR LIST OF RECYCLING DO'S AND DON'TS:

Do:

→ Wash and dry packaging before you put it into the recycling bin. Any leftover food will prevent all items from getting recycled.
→ Read the label and see exactly what part of the packaging can be recycled. Often you will need to separate the non-recyclable and recyclable bits.
→ Find out how your local recycling works. Do your research and you may be able to get extra bins for your area, such as a compost bin.

Don't:

→ Throw any old thing into the bin.
→ Put anything covered with food or grease into the bin.
→ Assume that everything goes into the same recycling bin.
→ Put the recycling in a plastic bag. It means that the recycling will automatically be taken to landfill instead – a waste of everyone's time.

Even better than recycling is reuse. Can you give something a second life? Just like with food waste, it's all about changing our mindset about what we've traditionally considered to be rubbish.

So you've got a leftover paper bag, can it be reused to take your sandwiches in for lunch? Or a yogurt pot – can you rinse and reuse it as a container for snacks?

→ Jars are a revelation here. Any time you use up something that has been in a jar, instead of throwing the jar away, give it a proper good wash with boiling water and use it to make dressings or delicious quick pickles.

SO, WHAT CAN BE RECYCLED?

Can:

→ **Clean glass** – bottles or jars of any colour
→ **Clean metals** – cans, foil, biscuit tins, aluminium trays
→ **Clean cardboard boxes** with their tape removed
→ **Clean hard plastics** – bottles, yogurt pots
→ **Clean paper** – non-shiny wrapping paper, brown paper, newspapers, envelopes with the plastic sheet removed

Hack: If you scrunch the paper and it doesn't spring back you can recycle it.

Can't:

→ **Coffee cups** – even though they look like paper the inside is coated in a thin layer of plastic
→ **Food-stained packaging** – pizza boxes, parchment paper, tin foil – they are a no go
→ **Soft plastics** – crisp packets, pasta packets, wrappers, clingfilm, straws, plastic bags. Most supermarkets now have a designated bin where you can recycle these materials
→ **Bubble wrap**
→ **Toothpaste tubes**
→ **Batteries**
→ **Nail varnish bottles**

Hack: If the plastic can be scrunched it can't be recycled.

You're now kitted out with all the tools to be as sustainable as humanly possible in the kitchen. This last chapter is all about practicality and helping you to become more sustainable in your everyday life.

We've got a bunch of hacks for when you're eating and drinking out, ideas for making holidays less wasteful and a bunch of recommendations for brilliant apps, resources and ways to get involved and keep you inspired.

From all of us at MOB Kitchen, thank you for joining us on this journey and being such environmentally conscious humans.

Ben x

TRASH TAKEAWAY

This one is tricky, we all love a takeaway, but every time you buy one that's a whole lot of single-use plastic. Many of the containers aren't recyclable, while the ones that are will end up in landfill (then in the ocean) unless cleaned properly.
Is a 20-minute delivery really worth up to 450 years of that plastic staying in the ocean?

Fear not, we've got some solutions:

1. **Eat in.** Get to know the restaurant behind your favourite pho takeaway and enjoy the atmosphere of eating in your top takeaway spot.

2. **Collection.** If you want to eat at home, take your own containers to the takeaway place for them to fill.

3. **Cook.** Ask yourself why you're ordering a takeaway. You now have the tools to make a meal out of virtually nothing and if it's something quick you're after then pasta or eggs is always a winning option.

4. **Reuse.** If takeaway is the only option, make sure you clean out and keep the takeaway boxes and reuse them as Tupperware.

HAPPY HOLIDAYS

Be it Christmas, Thanksgiving, Easter, Chinese New Year, Passover or Eid, holidays often mean feasting.

But, with feasting and lots of mouths to feed often comes a large amount of waste.

It's also much more stressful when you've got the whole MOB over, and so you have less time to think about all the ways you can be more sustainable. Here are some easy, simple tips to help you minimize excess waste:

1. **Eat from the freezer beforehand.** Save money before the holidays and save on food waste by having a freezer clear out so you've got the space to freeze leftovers.

2. **Make a shopping list.** Don't get distracted by holiday deals in the supermarket. Stick to only what you need to buy and try not to be enticed.

3. **Save on packaging.** Replace multi-pack treats with one large bag instead. Before buying, look to see what treats have the most recyclable packaging. If it's covered in soft plastic, it's a no. At Easter, look out for an eco egg or buy a chocolate bar instead.

4. **Go homemade where you can.** Simplify the food and all pitch in to help cook it so no one gets stressed out and you can ditch the shop-bought alternatives.

5. **Mix it up.** Does anyone in your house actually eat the Christmas pudding or are you buying it for the sake of it? Mix up traditions to suit what you actually like and avoid food waste.

OUT AND ABOUT

When you're out and about here are some ways to stay conscious:

1. Bring your **reusables**. We love coffee, but this is the year for change. If we haven't got our reusable cup with us, we aren't getting a coffee to take away. It's as simple as that.

2. What about **compostable cups**? Good question, but the bleak truth is that even if the cup is 100% compostable, unless it's being disposed of in an actual food waste compost bin (where have you seen one of those in public?), it is still going to end up in landfill, inside a plastic bag, unable to break down.

Make your own choices here. Be sure to ask the coffee shop what they do with the cups, but unless you can actually see them going into their own specific bin, chances are they won't be being composted.

The same goes for all compostable packaging. It's a brilliant idea but it's not the solution. Bring your reusables wherever you can.

3. Look out for **menu shout-outs**. If a restaurant is using ethical, good-quality ingredients the chances are that they are going to want to tell you about it.

LIFE HAPPENS

We are all on this journey together.

There will be times when you slip up. The fact that you care means that you are already starting to make a difference.

There's bound to be some changes that come easier to you than others. Remember, by being more sustainable we are choosing to say no to the norm.

These are the things we at MOB Kitchen are working on:

Ben: Stop wasting lone veggies that I've forgotten about in the fridge.
Alice: Buy less plastic-wrapped food.
Dan: Ditch the clingfilm once and for all.
Felix: Remember my reusables.
Soph: Use less water.
Polly: Be better with portion sizes.
Michael: Cut down on my meat consumption.

GET MORE INVOLVED

There are plenty of ways for you to get more involved if you want to.

Veganuary
Go vegan for a whole month in January. If you'd like to take part, sign up to the campaign for helpful recipes and tips or do it in your own time, whatever works.
https://uk.veganuary.com

Plastic Free July
Join the challenge to eliminate single-use plastic in July. The movement is full of ideas to help you get started and you'll become part of a growing community.
https://www.plasticfreejuly.org

Marches
Check online for the next climate change march. Every Friday you can strike under Greta Thunberg's Fridays For Future (FFF); find out locations or how to start your own strike online.
https://www.fridaysforfuture.org/

Join an organization
There are many different organizations out there for you to join. Do your research and see which one most aligns with your values. Thinking small scale is often good, you may find something going on in your local community that you can become a part of.

Sustainability club
Create your own club for sustainability. Each month you can set a challenge in the kitchen and then report back to the group on how it has gone. A great way of helping each other be better.

Have a conversation
Don't underestimate the power of talking MOB. Get into conversations about climate change and help educate others.

Put pressure on governments
Sign pledges and write emails to encourage legislative change.

RESOURCES

If you'd like more inspiration, here is a list of some of our favourite resources:

APPS

This is a shout-out to our top ten sustainability apps.

Olio
You can pick up and collect food or list food to be collected in your local area at a time convenient for you. We use it at MOB Kitchen for all our leftovers.

CozZo
Track what food you have at home and buy only what you need with this app. You can have up to 10 users per household so it's a great one to use with your flatmates to combat food waste.

WeTap
So, you've got your reusable water bottle but where do you fill it up? We Tap gives you the location of all public drinking fountains.

SDGs in Action
Keep up to date on the UN's 17 sustainability development goals with current news stories and ideas for how to get more involved if you're interested.

Love Food Hate Waste
This app offers simple tips to help you reduce your food waste. Great for when you are lacking in ideas and want to be re-inspired.

HowGood
A sustainable food directory that lists over 200,000 products. Simply scan the barcode to find out how ethical your food is before buying.

Too Good To Go
This app helps supermarkets, restaurants, cafés and bakeries combat food waste and means you get tasty food at a fraction of the price. At the end of each day unsold food is listed on the app ready for you to collect.

HappyCow: Find Vegan Food
Find vegan and vegetarian restaurants wherever you are with this handy app. Your best friend when travelling.

ShareTheMeal
An app from the UN World Food Programme. Buy a meal for someone less fortunate anywhere in the world with just one click. Being kind to others has never been so simple.

WEBSITES

The Intergovernmental Panel on Climate Change (IPCC) – https://www.ipcc.ch/

The World Counts
https://www.theworldcounts.com/

Global Climate Strike
https://globalclimatestrike.net/

Earth Overshoot Day
https://www.overshootday.org/

FridaysForFuture
https://www.fridaysforfuture.org/

Soil Association
https://www.soilassociation.org/

Food Forever
https://www.food4ever.org/

The Conversation
https://theconversation.com/uk

World Wildlife Fund
https://www.wwf.org.uk/

National Geographic
https://www.nationalgeographic.com/

Environmental Science
https://www.environmentalscience.org/

Food Miles https://www.foodmiles.com/

Marine Conservation Society
https://www.mcsuk.org/

Roundtable on Sustainable Palm Oil
https://rspo.org/

MBA Seafood Watch
https://www.seafoodwatch.org/

Love Food Hate Waste
https://lovefoodhatewaste.com/

Trash Plastic
https://www.trashplastic.com/

WRAP https://www.wrap.org.uk/

BOOKS

Will McCallum, *How To Give Up Plastic: Simple steps to living consciously on our blue planet*

Madeleine Olivia, *Minimal: How to simplify your life and live sustainably*

Michael Pollan, *The Omnivore's Dilemma: The search for a perfect meal in a fast-food world*

Erin Rhoads, *Waste Not: Make a Big Difference by Throwing Away Less*

Jonathan Safran Foer, *We are the Weather: Saving the Planet Begins at Breakfast*

Greta Thunberg, *No One Is Too Small to Make a Difference*

Giovanna Torrico & Amelia Wasiliev, *The Zero Waste Cookbook: 100 Recipes for Cooking Without Waste*

Bee Wilson, *The Way We Eat Now: strategies for eating in a world of change*

PODCASTS AND DOCUMENTARIES

TED Talks Daily podcast

The Food Programme podcast

Tristram Stuart: The global food waste scandal TED talk

Cowspiracy: The Sustainability Secret (Netflix)

The Game Changers (Netflix)

FOOTNOTES

Welcome

1. Oxfam International. (n.d.). '5 natural disasters that beg for climate action'. https://www.oxfam.org/en/5-natural-disasters-beg-climate-action

2. Buis, A. (19 June 2019). 'A degree of concern: why global temperatures matter'. NASA: Climate Change and Global Warming. https://climate.nasa.gov/news/2865/a-degree-of-concern-why-global-temperatures-matter/

3. Leahy, S. (7 May 2019). 'One million species at risk of extinction, UN report warns'. National Geographic. https://www.nationalgeographic.co.uk/environment/2019/05/one-million-species-risk-extinction-un-report-warns

4. Ritchie, H. (6 November 2019). 'Food production is responsible for one-quarter of the world's greenhouse gas emissions'. Our World in Data. https://ourworldindata.org/food-ghg-emissions

5. Shukman, D. (2 July 2019). "Football pitch' of Amazon forest lost every minute'. BBC News. https://www.bbc.co.uk/news/science-environment-48827490

6. Harvey, C. (14 June 2018). 'Antarctic melt rate has tripled in the last 25 years'. Scientific American. https://www.scientificamerican.com/article/antarctic-melt-rate-has-tripled-in-the-last-25-years/

7. Majid, A. (1 May 2018). 'WHO reveals 7 million die from pollution each year in latest global air quality figures'. The Telegraph. https://www.telegraph.co.uk/global-health/climate-and-people/estimates-7-million-die-pollution-year-reveals-latest-global/

8. Earth Overshoot Day marks the date when humanity's demand for ecological resources and services in a given year exceeds what Earth can regenerate in that year. https://www.overshootday.org/

9. Higgins, E. (27 September 2019). 'With over 6 million people worldwide, climate strikes largest coordinated global uprising since Iraq War protests'. Common Dreams. https://www.commondreams.org/news/2019/09/27/over-6-million-people-worldwide-climate-strikes-largest-coordinated-global-uprising

10. Rivera, L. (21 January 2020). Sustainability journalist for The Independent talking at a panel discussion.

11. Alberro, H. (28 December 2019). 'Climate change: six positive news stories from 2019: Costa Rica offers us a viable climate future'. https://theconversation.com/climate-change-six-positive-news-stories-from-2019-129100

12. Maslin, M. (28 December 2019). 'Climate change: six positive news stories from 2019: Young people are on the march! https://theconversation.com/climate-change-six-positive-news-stories-from-2019-129100

13. Wikipedia. (referenced 1 April 2020). Category: Climate change organizations. https://en.wikipedia.org/wiki/Category:Climate_change_organizations

14. The European Green Deal is a roadmap for making the EU's economy sustainable. European Commission. https://ec.europa.eu/info/strategy/priorities-2019-2024/european-green-deal_en

15. Catholic Climate Covenant. (n.d.). 'Pope Francis encyclical and climate change'. https://catholicclimatecovenant.org/encyclical

16. BBC News. (29 July 2019). 'Ethiopia breaks tree-planting record to tackle climate change'. https://www.bbc.co.uk/news/world-africa-49151523

Time for Change

17. Food and Agricultural Organization of the United Nations. (19 February 2015). '2015 international year of soils'. http://www.fao.org/soils-2015/news/news-detail/en/c/277682/

18. Warren, J. (15 January 2016). 'Why do we consume only a tiny fraction of the world's edible plants?'. The World Economic Forum. https://www.weforum.org/agenda/2016/01/why-do-we-consume-only-a-tiny-fraction-of-the-world-s-edible-plants

Diet

19. Gallagher, J. (17 January 2019). 'A bit of meat, a lot of veg – the flexitarian diet to feed 10bn'. BBC News. https://www.bbc.co.uk/news/health-46865204

20. J. Poore, J and Nemecek, T. (1 June 2018). 'Reducing food's environmental impact through producers and consumers'. Figures and Data. Science Magazine. Vol. 360, Issue 6392, pp. 987–992. https://science.sciencemag.org/content/360/6392/987/tab-figures-data and Moskin, J et al. (30 April 2019). 'Your questions about food and climate change answered'. The New York Times. https://www.nytimes.com/interactive/2019/04/30/dining/climate-change-food-eating-habits.html

21. Mbow, C and Rosenzweig, C et al. (11 August 2019). 'Food Security'. Climate Change and Land: an IPCC special report. Chapter 5, 5.4.3 https://www.ipcc.ch/site/assets/uploads/sites/4/2019/11/08_Chapter-5.pdf

22. Mbow, C and Rosenzweig, C et al. (11 August 2019). 'Food Security'. Climate Change and Land: an IPCC special report. Chapter 5, 5.4.6 https://www.ipcc.ch/site/assets/uploads/sites/4/2019/11/08_Chapter-5.pdf

23. Hill, S and Annesi, K. (n.d.). 'Climate change and our food system'. Plant Proof. https://plantproof.com/animal-vs-plant-agriculture-in-the-era-of-climate-change/

24. Calculated on Sainsbury's groceries nutrition values Sainsburys.co.uk

Sustainable Food Choices

25. SBS News. (3 September 2019). 'Food miles: myth or fact?'. https://www.sbs.com.au/news/food-miles-myth-or-fact

26. Weber, C. L and Matthews, H. S. (16 April 2008). 'Food-miles and the relative climate impacts of food choices in the United States'. Environ. Sci. Technol. 42, 10, 3508-3513. https://pubs.acs.org/doi/full/10.1021/es702969f

27. ETA Services. (n.d.). 'Food miles'. https://www.eta.co.uk/environmental-info/food-miles/ and Wilson, L . (n.d.). 'The tricky truth about food miles'. Shrink that Footprint. http://shrinkthatfootprint.com/food-miles

28. World Wildlife Fund. (n.d.). 'Which everyday products contain palm oil'. https://www.worldwildlife.org/pages/which-everyday-products-contain-palm-oil

29. World Wildlife Fund. (17 January 2020). '8 things to know about palm oil'. https://www.wwf.org.uk/updates/8-things-know-about-palm-oil

30. Murphy, D. J. (30 June 2015).'Palm oil: scourge of the earth, or wonder crop?'. The Conversation. https://theconversation.com/palm-oil-scourge-of-the-earth-or-wonder-crop-42165

31. World Wildlife Fund. (n.d.). 'Sustainable

agriculture: soy'. https://www.worldwildlife.org/industries/soy

32. Rizzo, G and Baroni, L. (5 January 2018). 'Soy, Soy Foods and Their Role in Vegetarian Diets'. *Nutrients*. 10(1): 43. https://www.ncbi.nlm.nih.gov/pmc/articles/PMC5793271/

33. Greenpeace. (n.d.). 'The challenges we face: soya'. https://www.greenpeace.org.uk/challenges/soya/

34. FAIRR. (n.d.). 'Amazon Soy Moratorium'. https://www.fairr.org/engagements/amazon-soy-moratorium/

35. Viva! (n.d.). 'Is soya destroying the planet?'. https://www.viva.org.uk/soya-planet and World Wildlife Fund. (14 December 2016). 'The story of soy'. https://www.worldwildlife.org/stories/the-story-of-soy

36. Moskin, J et al. (30 April 2019). 'Your questions about food and climate change answered'. *The New York Times*. https://www.nytimes.com/interactive/2019/04/30/dining/climate-change-food-eating-habits.html

37. Nair, K. (16 January 2020). 'The rise of blockchain in seafood traceability'. Supply Chain Brain. https://www.supplychainbrain.com/blogs/1-think-tank/post/30703-the-rise-of-blockchain-in-seafood-traceability

38. Hill, J. (n.d.). 'Environmental consequences of fishing practices'. Environmental science. https://www.environmentalscience.org/environmental-consequences-fishing-practices#_ENREF_17

39. NOAA Fisheries. (n.d.). 'Ending overfishing through annual catch limits'. https://www.fisheries.noaa.gov/national/rules-and-regulations/ending-overfishing-through-annual-catch-limits and European Commission. (n.d.). 'The common fisheries policy' (CFP). https://ec.europa.eu/fisheries/cfp_en

40. Cohen, J. (15 January 2020). 'Farmed fish vs wild fish: is one always better than the other?'. Chowhound. https://www.chowhound.com/food-news/190114/farmed-fish-vs-wild-fish-what-is-the-difference/

41. Barlow, J. and Cameron, G.A. (26 August 2006). 'Field experiments show that acoustic pingers reduce marine mammal bycatch in the California drift gill net fishery'. *Marine Mammal Science*, 19(2): pp. 265–283. https://onlinelibrary.wiley.com/doi/abs/10.1111/j.1748-7692.2003.tb01108.x and Lewison, R.L. et al. (4 August 2003). 'The impact of turtle excluder devices and fisheries closures on loggerhead and Kemp's ridley strandings in the western Gulf of Mexico'. *Conservation Biology*. 17(4): pp. 10891097. http://www.conservationecologylab.com/uploads/1/9/7/6/19763887/lewison_et_al_2003.pdf

42. Greenaway, T. (4 January 2012). 'Is your all-you-can-eat shrimp killing the mangroves?'. Grist. https://grist.org/food/2012-01-03-is-your-all-you-can-eat-shrimp-killing-the-mangroves/ and

43. Fredriksson, O. (6 December 2019). 'Fish farming: opinion on aquaculture'. Food Unfolded. https://www.foodunfolded.com/opinion/fish-farming-opinion-on-aquaculture?gclid=EAIaIQobChMIzd7a_-Kt5wIVw7TtCh1nYgGYEAAYASAAEgIHx_D_BwE

44. Scottish Government. (n.d.). 'Fish health inspectorate'. https://www2.gov.scot/Topics/marine/Fish-Shellfish/FHI

45. Rubicon Resources. (29 January 2018). '3 promising alternative feeds for aquaculture'. https://medium.com/sustainable-seafood/3-promising-alternative-feeds-for-aquaculture-2742c011e3cc

46. Marine Conservation Society. https://www.mcsuk.org/who-we-are/

47. Seafood Watch https://www.seafoodwatch.org/

48. Marine Stewardship Council https://www.msc.org/uk

49. Aquaculture Stewardship Council https://www.asc-aqua.org/

50. Soil Association. https://www.soilassociation.org/organic-living/

51. Nicholson, R. (28 February 2017). 'What does free-range actually mean?'. *The Guardian*. https://www.theguardian.com/lifeandstyle/shortcuts/2017/feb/28/what-does-free-range-actually-mean-its-complicated

52. Fairtrade https://www.fairtrade.org.uk/

53. Fairtrade premium https://www.fairtrade.org.uk/What-is-Fairtrade/What-Fairtrade-does/Fairtrade-Premium

54. RSPCA https://www.rspcaassured.org.uk/

55. Red Tractor https://www.redtractor.org.uk/

56. British Lion https://www.egginfo.co.uk/british-lion-eggs

57. Eshel, G et al. (19 August 2014). 'Land, irrigation water, greenhouse gas, and reactive nitrogen burdens of meat, eggs, and dairy production in the United States'. PNAS 111 (33) 11996-12001 https://www.pnas.org/content/111/33/11996 and Healthyish. (30 January 2020). 'The healthyish guide to eating for the planet... without stressing out'. Bon Appetit. https://www.bonappetit.com/gallery/eating-for-the-climate

58. Westcott, R. (19 June 2018). 'Top of the swaps: five fish to avoid & what to eat instead'. The Jellied Eel. https://www.sustainweb.org/jelliedeel/articles/jun18_topoftheswaps/ and Healthyish. (30 January 2020). 'The healthyish guide to eating for the planet... without stressing out'. Bon Appetit. https://www.bonappetit.com/gallery/eating-for-the-climate

59. Saner, E. (21 October 2015). 'Almond milk: quite good for you – very bad for the planet'. *The Guardian*. https://www.theguardian.com/lifeandstyle/shortcuts/2015/oct/21/almond-milk-quite-good-for-you-very-bad-for-the-planet

60. Arbor Day Foundation. (n.d.). 'The promise of sustainable agriculture: environmental & agricultural benefits'. https://www.arborday.org/programs/hazelnuts/consortium/agriculture.cfm

Food Waste

61. Food and Agricultural Organization of the United Nations. (n.d.). 'Food loss and food waste'. http://www.fao.org/food-loss-and-food-waste/en/

62. WRAP. (January 2020). 'Food surplus and waste in the UK – key facts'. https://wrap.org.uk/sites/files/wrap/Food %20surplus_and_waste_in_the_UK_key_facts_Jan_2020.pdf

63. Brackett, B. (2 February 2016). 'The difference between use-by, sell-by and best-by dates'. Institute of Food Technologists. https://www.ift.org/career-development/learn-about-food-science/food-facts/the-difference-between-useby-sellby-and-bestby-dates

64. Olsson, K. Karlstad University. (30 January 2018). 'Bananas are some of the worst food waste culprits, study shows'. Science Daily. https://www.sciencedaily.com/releases/2018/01/180130091407.htm

65. WRAP. (7 November 2013). 'Use your loaf and save billions'. https://www.wrap.org.uk/content/use-your-loaf-and-save-billions

66. Ziane, L. (13 February 2020). 'Bread waste is off the scale'. Toast. https://www.toastale.com/bread-waste/

67. Smithers, R. (November 2017). 'Nearly half of all fresh potatoes thrown away daily by UK households'. *The Guardian*. https://www.theguardian.com/environment/2017/nov/08/nearly-half-of-all-fresh-potatoes-thrown-away-daily-by-uk-households

Water

68. Spector, D. (14 February 2016). 'How long a person can survive without water'. *The Independent* https://www.independent.co.uk/life-style/health-and-families/health-news/how-long-a-person-can-survive-without-water-a6873341.html

69. Ritchie, H and Roser, M. (November 2019). 'Clean water'. Our World in Data. https://ourworldindata.org/water-access

70. United Nations News Centre. (12 July 2017). 'Billions around the world lack safe water, proper sanitation facilities, reveals UN report'. https://www.un.org/sustainabledevelopment/blog/2017/07/billions-around-the-world-lack-safe-water-proper-sanitation-facilities-reveals-un-report/

71. United Nations Water. (2017). 'Water and gender'. https://www.unwater.org/water-facts/gender/

72. United Nations Water. (2017). 'Water and gender'. https://www.unwater.org/water-facts/gender/

73. World Health Organization and United Nations Children's Fund. (2017). 'Progress on drinking water, sanitation and hygiene'. p.29. https://apps.who.int/iris/bitstream/handle/10665/258617/9789241512893-engdf.jsessionid=330207F9B3CAA25D492D57534953FE93?sequence=1.

74. UNICEF Publications. (March 2017). 'Thirsting for a Future: water and children in a changing climate'. p.19 https://www.unicef.org/publications/index_95074.html and Schulte, P. (4 February 2014). 'Defining water scarcity, water stress and water risk'. Pacific Institute. https://pacinst.org/water-definitions/ and United Nations Water. (n.d.) 'Water scarcity'. https://www.unwater.org/water-facts/scarcity/

75. Coping with Water Scarcity (2007) p.9 http://www.fao.org/3/a-aq444e.pdf

76. Be Fresh. (12 April 2018). 'How much water does it take to produce your fave foods?'. https://www.befresh.ca/blog-how-much-water/, https://thewaterweeat.com/ and *The Guardian* data blog. (January 2013). 'How much water is needed to produce food and how much do we waste?'. https://www.theguardian.com/news/datablog/2013/jan/10/how-much-water-food-production-waste#data

Plastic

77. Weikle, B. (21 May 2018). 'Plastics and

other garbage found in ocean trench nearly 11 kilometres below surface'. CBC News. https://www.cbc.ca/news/technology/plastic-deep-sea-debris-ocean-trench-1.4667038

78. McCallum, W. (24 May 2018). *How to give up plastic*. p.35

79. McCallum, W. (24 May 2018). *How to give up plastic*. p.47

80. McCallum, W. (24 May 2018). *How to give up plastic*. p.47 and Plastic Oceans. (n.d.). 'The Facts'. https://plasticoceans.org/the-facts/

81. Plastic Oceans. (n.d.). 'Plastic pollution facts'. https://plasticoceans.uk/the-facts-plastic-pollution-2/

82. *National Geographic* encyclopaedic entry. 'The great Pacific garbage patch'. https://www.nationalgeographic.org/encyclopedia/great-pacific-garbage-patch/

83. The Ellen MacArthur Foundation. (13 December 2017). 'The new plastics economy: rethinking the future of plastics & catalysing action'. https://www.ellenmacarthurfoundation.org/publications/the-new-plastics-economy-rethinking-the-future-of-plastics-catalysing-action

84. Parker, L. (7 June 2019). 'The world's plastic pollution crisis explained'. *National Geographic*. https://www.nationalgeographic.com/environment/habitats/plastic-pollution/

85. Wright, M et al. (10 January 2018). 'The stark truth about how long your plastic footprint will last on the planet'. *The Telegraph*. https://www.telegraph.co.uk/news/2018/01/10/stark-truth-long-plastic-footprint-will-last-planet/

86. World Health Organization. (2019). 'Microplastics in drinking water'. https://apps.who.int/iris/bitstream/handle/10665/326499/9789241516198-eng.pdf?ua=1 and Sea Grant. 'Marine debris is everyone's problem'. https://seagrant.whoi.edu/wp-content/uploads/sites/106/2015/04/Marine-Debris-Poster_FINAL.pdf